Plane Waves
and Spherical Means

Fritz John

Plane Waves and Spherical Means

Applied to Partial Differential Equations

Springer Science+Business Media, LLC

Fritz John

Courant Institute of Mathematical
Sciences
New York University
251 Mercer Street
New York, New York 10012
USA

AMS Classification: 35A22, 35A25, 35A30, 35L05, 35D05

With 10 figures

Publisher of Original Edition: Interscience Publishers, Inc., New York, 1955.

9 8 7 6 5 4 3 2 1
ISBN 978-0-387-90565-5 ISBN 978-1-4613-9453-2 (eBook)
DOI 10.1007/978-1-4613-9453-2

FOREWORD

The author would like to acknowledge his obligation to all his colleagues and friends at the Institute of Mathematical Sciences of New York University for their stimulation and criticism which have contributed to the writing of this tract. The author also wishes to thank Aughtum S. Howard for permission to include results from her unpublished dissertation, Larkin Joyner for drawing the figures, Interscience Publishers for their cooperation and support, and particularly Lipman Bers, who suggested the publication in its present form.

New Rochelle
September, 1955

FRITZ JOHN

CONTENTS

CHAPTER V

The Theorems ot Asgeirsson and Howard

CHAPTER VI

Determination of a Function from its Integrals over Spheres of a Fixed Radius

CHAPTER VII

Differentiability Properties of Solutions of Elliptic Systems

CHAPTER VIII

Regularity Properties for Integrals of Solutions over Time-like Lines

INTRODUCTION

This tract contains a somewhat heterogeneous collection of results on partial differential equations. The unifying element is the use of certain elementary identities for plane and spherical integrals of an arbitrary function. It is the aim of the author to show how a variety of results on fairly general differential equations follows from those identities.

The use of ordinary euclidean planes and spheres in connection with general differential equations represents a departure from the idea that it is best to work with geometric entities like characteristic conoids, which are associated in an invariant manner with the differential equation. It is probably true that the finer structure of the solutions is only brought out by using an invariant approach, adjusted to the individual equation. On the other hand experience shows that many results have been obtained more easily by employing cruder unspecific tools, such as power series, Fourier integrals, finite difference approximations, or L^2-norms. The characteristic properties of the individual equation then enter only through *inequalities* instead of equations, and the considerable difficulties inherent in the use of singular integrals over characteristic conoids are avoided. In the same way it will be seen here that integrals over ordinary spheres and planes can be used to advantage even for equations that are not related to the ordinary euclidean metric. In such cases the use of these euclidean objects will introduce certain artificial (non-invariant) features. This is compensated for by the simplicity and symmetries of ordinary planes and spheres compared with the corresponding objects (if any) in whatever geometry might be associated *naturally* with the differential equation.

Most of the results given here can be found scattered in the literature, though possibly with different degrees of generality. A conscientious effort has been made to give appropriate credit to other authors and to provide references to related material.

No attempt has been made to give a historical survey and to decide more intricate questions of priority. This would represent a formidable task in the field of partial differential equations, where the actual results of various workers often do not differ as much (and perhaps are not of as much interest) as their emphasis on some specific unifying point of view. The results given here have been selected so as to demonstrate best the usefulness of plane waves and spherical means. As far as possible the treatment has been made elementary and self-contained. It is clear that this imposes severe restrictions on the choice of topics and precludes an exhaustive treatment of any one subject.

The basic identities applied in this monograph are contained in Chapters I and IV. Chapter I deals with the *decomposition of arbitrary functions into functions of the type of plane waves*, i.e. into functions that have parallel planes as level surfaces. Fourier analysis provides one such decomposition, namely into *plane waves of exponential type*. For many applications the exponential character is not essential, and more elementary ways of decomposing a function into plane waves can be used. One such method of decomposition is given here. It consists of expressing a function by spherical means of integrals of the function over hyper-planes, and is due to J. Radon [1].[1] The resulting formulae are closely related to those giving the solution of the initial value problem of the wave equation.[2] Their connection with more general hyperbolic equations with constant coefficients was indicated by G. Herglotz [3], p. 18. This type of decomposition of a function into plane waves could be called the *Radon transform* in contrast to the *Fourier* transform.

Chapter II brings as the first application of the Radon transformation the solution of the *initial value problem for homogeneous hyperbolic equations with constant coefficients.* ("Homogeneous" here refers to the assumption that all derivatives occurring in the equation are of the same order.[3]) The formulae we arrive

[1] Numbers in brackets refer to the bibliography at the end of the tract.
[2] See Mader [1].
[3] For a complete discussion (including inhomogeneous equations) by means of symbolic calculus the reader is referred to Leray [1], [2].

at go back to Herglotz [4] (though with some restrictions of generality) and were extended by Bureau, Gårding, Leray and Petrovskii. They can be obtained in principle by starting from Cauchy's solution [5] by Fourier integrals and by "evaluating" the kernel arising from interchange of integrations. This method of approach is bound to run into convergence difficulties, especially in the case, where the order of the equation is less than the number of dimensions. In the latter case there just does not exist an "integral representation" of the solution in terms of the data in the ordinary sense, since the solution does not depend *continuously* on the data, if the *maximum norm* is adopted. All one has a right to expect is that the solution of the initial value problem for a hyperbolic equation depends continuously on the initial data *and* on a finite number of their derivatives. [6] If an integral representation is to be used, it has to be interpreted in some generalized sense, say as "finite" or "logarithmic" part of an improper integral, as in Hadamard's theory of second order equations and in the work of Bureau, [7] or following M. Riesz [1] by analytic continuation of proper integrals, or as a distribution in the sense of L. Schwartz [2]. All these generalized integral representations can be made "concrete" in the form of derivatives of ordinary integrals, though the transition may require rather unwieldy computations with singular integrals. In contrast to that the method employed here (for homogeneous equations) avoids all convergence difficulties by working with the simpler Radon decomposition into plane waves instead of the Fourier integrals. The solution is then obtained immediately in the form of an iterated Laplacean applied to a perfectly regular integral operator acting on the initial data. It is only when one attempts to simplify the expression by carrying out some of the differentiations explicitly that the classical difficulties re-appear. [7a]

[4] See Herglotz [1], [2]. Herglotz also gave an exposition of the subject in his course on "Mechanik der Kontinua," Göttingen, 1931 (see [3]).

[5] See Cauchy [1], Courant-Hilbert [1], vol. II, ch. III, Bureau [9].

[6] See Hadamard [1], Gårding [1].

[7] See Bureau [3], [4], [9].

[7a] For a related procedure for general hyperbolic systems with constant coefficients see R. Courant and A. Lax [1].

Chapter III gives the construction of the *fundamental solution* for a linear elliptic equation, and more generally for a linear elliptic system, with analytic coefficients. The problem amounts to finding a solution of the symbolic equation

$$L[u] = \delta,$$

where δ is the Dirac function. The method of solution given here amounts to decomposing the Dirac function into plane waves and thus to reducing the problem to that of finding a solution of $L[u] = f$, where f is a plane wave function. The latter problem in turn can be solved as a consequence of the theorem of Cauchy and Kowalewski. In this way a fundamental solution in the small is obtained in a form for which it is easy to analyze the nature of the singularity to any order of magnitude desired. [8] For equations with constant coefficients one finds explicit expressions for the fundamental solution, which only involve quadratures. This special case is of importance, because it furnishes the *parametrix* solutions that can be used for general linear elliptic equations with non-analytic coefficients. [9]

Chapter IV derives *expressions for an arbitrary function f in terms of spherical integrals of f.* For the applications it is important that the radii of the spheres occurring in those expressions are bounded away from zero. These formulae form the principal tool used in the remaining chapters. They can be looked at as generalizations of the formulae of Chapter I, which express f in terms of its integrals over planes. The resulting formulae for f in terms of its spherical means are not particularly elegant. Fortunately it is only the general form of the expression that matters for the later applications. The identities in Chapter IV are closely related to Huygens' principle for the wave equation and to certain identities for Bessel functions. They can also be looked at as the analytic counterpart of the geometrical observation that spherical shells can be swept out by spheres in two different ways

[8] For the case of equations of second order the fundamental solution had been constructed by Hadamard [1]; book II, Ch. III). See also Thomas and Titt [1], Bureau [2], Miranda [1].

[9] See E. E. Levi [1], John [7], pp. 155—162.

(see Fig. 8), just as the identities of Chapter I are connected with the fact that the exterior of a sphere can be swept out by planes.[10]

Chapter V brings the *identity of Asgeirsson* together with a somewhat more general *identity due to A. Howard.* The latter identity illuminates Asgeirsson's theorem by relating it to the geometry of linear families of quadrics in tangential coordinates. The theorem of Mrs. Howard exceeds the bounds set to this monograph by its title, in so far as it deals with *ellipsoidal* means instead of *spherical* ones. It has the interesting application that it permits to transform a homogeneous differential equation of order $2m$ with constant coefficients into a similar equation of order m in more independent variables.

Chapter VI deals mostly with the problem of determining a function from its integrals over spheres of radius 1. The problem can be solved on the basis of the identities of Chapter IV. The solution by decomposition into plane waves of more general problems for "mean-periodic" functions is indicated.

Chapter VII gives the main application of the identities on spherical means derived in Chapter IV. It presents proofs for the differentiability of solutions of linear or non-linear elliptic equations or of systems of equations, provided the coefficients are sufficiently regular. It also contains a proof for the analyticity of solutions of linear elliptic equations with analytic coefficients. (Another proof is implicit in the construction of analytic fundamental solutions of such equations in Chapter III.)

Chapter VIII contains an extension of the results of Chapter VII to linear *non-elliptic* equations. Here not the regularity of the *solutions* but of certain *integral transforms of the solution* is established. More precisely integrals of a solution over a family of *time-like* curves with common endpoints are shown to depend as regular on the parameter distinguishing the members of the family as the coefficients of the differential equation and the regularity assumptions on the curves of the family permit.

[10] Similarly Asgeirsson's theorem in $2+2$ dimensions corresponds to the geometrical fact that a hyperboloid of one sheet is covered by straight lines in two distinct ways. See John [6].

CHAPTER I

Decomposition of an Arbitrary Function into Plane Waves

Explanation of notation

In what follows the letters x, y, z, X, Y, Z, ξ, η, ζ will always stand for the *vectors* $(x_1, \ldots, x_n)$, $(y_1, \ldots, y_n), \ldots, (\zeta_1, \ldots, \zeta_n)$ in n-dimensional space where $n \geqq 2$. All other letters will stand for *scalar* variables. The *scalar product* $\sum_{i=1}^{n} x_i y_i$ of the vectors x and y will be denoted by $x \cdot y$, the length $(x \cdot x)^{1/2}$ of the vector x by $|x|$. The volume element $dx_1 \ldots dx_n$ will be abbreviated to dx, while dS_x will denote the surface element of a hyper-surface in n-dimensional space. The spherical surface of radius 1 about the origin in x-space will be denoted by Ω_x, its surface element by $d\omega_x$, its total surface measure by ω_n. The volume of the unit-sphere in n-space is then $(1/n)\omega_n$. Integrations are carried out over the whole range of a variable, unless other limits are indicated.

The spherical mean of a function of a single coordinate

Let $g(s)$ be a continuous function of the scalar variable s. Denoting by y a fixed vector, we have in $g(y \cdot x)$ a function of $x = (x_1, \ldots, x_n)$, which is constant along the hyperplanes perpendicular to the direction of y; (such a function will be called a "plane wave" function). We form the integral of $g(y \cdot x)$ over the solid sphere of radius r about the origin by decomposing the sphere into plane sections perpendicular to the y-direction. On the plane $y \cdot x = |y| p$ of distance $|p|$ from the origin the function $g(x \cdot y)$ has the value $g(|y| p)$. The $(n-1)$-dimensional intersection of that plane with the sphere has the volume (see Fig. 1)

$$\frac{\omega_{n-1}}{n-1}(r^2-p^2)^{(n-1)/2}.$$

It follows that

$$(1.1)\qquad \int_{|x|<r} g(y\cdot x)dx = \frac{\omega_{n-1}}{n-1}\int_{-r}^{+r}(r^2-p^2)^{\frac{n-1}{2}}g(|y|p)dp.$$

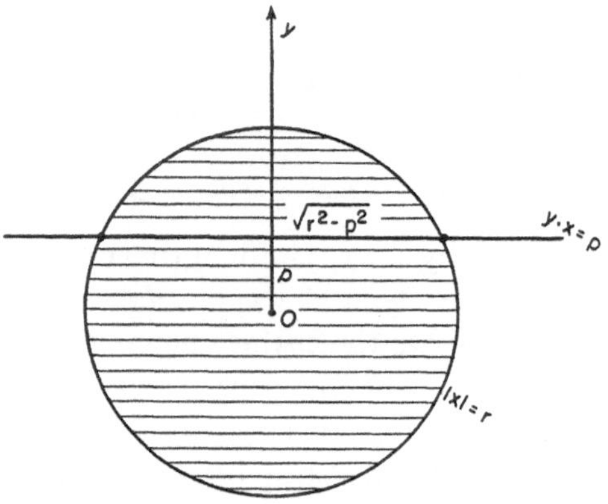

Figure 1

Differentiating with respect to r and putting $r=1$ we obtain the fundamental identity

$$(1.2)\qquad \int_{\Omega_x} g(y\cdot x)d\omega_x = \omega_{n-1}\int_{-1}^{+1}(1-p^2)^{(n-3)/2}g(|y|p)dp = \omega_n h(|y|)$$

for the integral of a plane wave function over the unitsphere, valid for $n \geqq 2$. [Here h is defined by (1.2).]

For $g(s) = \text{const.} = 1$ we have $h = 1$, and (1.2) yields the recursion formula

$$(1.3)\qquad \frac{\omega_n}{\omega_{n-1}} = \int_{-1}^{1}(1-p^2)^{(n-3)/2}dp = \frac{\Gamma\left(\frac{n-1}{2}\right)\Gamma\left(\frac{1}{2}\right)}{\Gamma\left(\frac{n}{2}\right)}.$$

From this formula one derives the well known value

$$\omega_n = \frac{2\sqrt{\pi}^n}{\Gamma\left(\frac{n}{2}\right)} \tag{1.4}$$

for the surface area of the unit sphere in n-space.[11]

For $g(s) = e^{is}$ we find

$$h(s) = \frac{\omega_{n-1}}{\omega_n} \int_{-1}^{1} (1 - p^2)^{\frac{n-3}{2}} e^{isp}\, dp = \frac{2^\nu \Gamma(\nu + 1)}{s^\nu} J_\nu(s), \tag{1.5}$$

where J_ν is the Bessel function of index $\nu = (n - 2)/2$.[12]

Taking $g(s) = |s|^k$ and $g(s) = |s|^k \log |s|$ in (1.2) yields respectively the identities

$$\int_{\Omega_x} |y \cdot x|^k \, d\omega_x = \frac{2\sqrt{\pi}^{n-1} \Gamma\left(\frac{k+1}{2}\right)}{\Gamma\left(\frac{n+k}{2}\right)} |y|^k \tag{1.6}$$

$$\int_{\Omega_x} |y \cdot x|^k \log |y \cdot x| \, d\omega_k = \frac{2\sqrt{\pi}^{n-1} \Gamma\left(\frac{k+1}{2}\right)}{\Gamma\left(\frac{n+k}{2}\right)} |y|^k (\log|y| + c_{nk}) \tag{1.7}$$

with a certain numerical constant c_{nk}. Formulae (1.6), (1.7) have been derived under the assumption that $g(s)$ is continuous, and hence that $k > 0$; they obviously then also hold for $k = 0$. They form the basis of the decomposition of an arbitrary function into plane waves, discussed in the next section.

Representation of a function by its plane integrals

We consider an arbitrary function $f(x)$ of class C_1, which

[11] See Courant-Hilbert [1], vol. II, p. 223.

[12] This is essentially Poisson's representation of the Bessel functions. See Magnus-Oberhettinger [1], p. 26, § 5.

vanishes outside a bounded set.[13] Then

$$u(z) = \int f(y) \frac{|y - z|^{2-n}}{(2-n)\omega_n} dy \tag{1.8}$$

is a function of z of class C_2, which satisfies Poisson's differential equation

$$\Delta_z u(z) = f(z) \tag{1.9}$$

where Δ_z denotes the Laplacean with respect to the variables $z_1, \ldots, z_n$. (For $n = 2$ the kernel has to be replaced by $(1/2\pi) \log |y - z|$.) For the proof[14] of (1.9) we observe that

$$\begin{aligned}
\Delta_z u &= \frac{-1}{\omega_n} \sum_i \frac{\partial}{\partial z_i} \int f(y)\,(y_i - z_i)\,|y - z|^{-n} dy \\
&= \frac{-1}{\omega_n} \sum_i \frac{\partial}{\partial z_i} \int f(y + z) y_i\,|y|^{-n} dy \\
&= -\frac{1}{\omega_n} \sum_i \int f_{y_i}(y + z) y_i\,|y|^{-n} dy \\
&= -\frac{1}{\omega_n} \lim_{r \to 0} \sum_i \int_{|y|>r} f_{y_i}(y + z) y_i\,|y|^{-n} dy \\
&= -\frac{1}{\omega_n} \lim_{r \to 0} \sum_i \left[\int_{|y|=r} \frac{-y_i^2}{r} |y|^{-n} f(y+z)\, dS_y - \int_{|y|>r} f(y+z) \frac{\partial}{\partial y_i}(y_i |y|^{-n}) dy \right] \\
&= \frac{1}{\omega_n} \lim_{r \to 0} r^{1-n} \int_{|y|=r} f(y + z)\, dS_y = f(z)
\end{aligned}$$

The proof of Poisson's equation has been given here in detail, because of its fundamental importance for what follows, since most of the differentiations of singular integrals that will have to be carried out will be reduced to this one formula. It may be mentioned that the same equation can be established under the milder assumption that $f(x)$ satisfies a Hölder condition.[15]

[13] $f(x)$ is of class C_m, if f and all its derivatives of orders $\leq m$ are continuous.
[14] See Courant-Hilbert [1], vol. II, p. 228.
[15] See Kellog [1], p. 156.

We now take for even n identity (1.7), for odd n identity (1.6), replace y by $y - z$, multiply with $f(y)$ and integrate over all y. (We still assume that f is of class C_1 and vanishes outside a bounded set.) We choose for k a non-negative integer such that $n + k$ is an even number, and apply the operator Δ_z to the resulting equation $(n + k)/2$-times. Observing that

$$\Delta_z \mid y - z \mid^k = k(k + n - 2) \mid y - z \mid^{k-2}$$

we find respectively for odd and even $n > 2$

$$(1.9a) \quad (\Delta_z)^{(n+k-2)/2} | y - z |^k = \frac{2^{n+k-1} \Gamma\left(\frac{k+2}{2}\right) \Gamma\left(\frac{k+n}{2}\right) \Gamma\left(\frac{n}{2}\right)}{(2 - n)\pi} (-1)^{(n-1)/2} | y - z |^{2-n}$$

$$(1.9b) \quad (\Delta_z)^{(n+k-2)/2} | y - z |^k \log | y - z | = \frac{2^{n+k-2} \Gamma\left(\frac{k+2}{2}\right) \Gamma\left(\frac{k+n}{2}\right) \Gamma\left(\frac{n}{2}\right)}{2 - n} (- 1)^{(n-2)/2} | y - z |^{2-n}$$

Hence from (1.6), (1.7), (1.9)

$$(1.10) \quad (\Delta_z)^{(n+k)/2} \int \left(\int_{\Omega_x} f(y) \mid (y - z) \cdot x \mid^k d\omega_x \right) dy = 4(2\pi i)^{n-1} k! f(z)$$

for odd n and $k = 1, 3, 5, \ldots$

$$(1.11) \quad (\Delta_z)^{(n+k)/2} \int \left(\int_{\Omega_x} f(y) ((y - z) \cdot x)^k \log \mid (y - z) \cdot x \mid d\omega_x \right) dy = - (2\pi i)^n k! f(z)$$

for even n and $k = 0, 2, 4, \ldots$ (also for $n = 2$).

We can formally combine these formulae for even and odd n into

$$(1.11a) \quad f(z) = (\Delta_z)^{(n+k)/2} \mathscr{Re} \left[- \frac{1}{k!(2\pi i)^n} \int \left(\int_{\Omega_x} f(y) [(y - z) \cdot x]^k \times \log \left[\frac{1}{i} (y - z) \cdot x \right] d\omega_x \right) dy \right]$$

where $\log s$ denotes the principal branch of that function defined in the complex s-plane slit along the negative real axis.

Formulae (1.10), (1.11) represent a solution of the problem of obtaining a function $f(z)$ as a linear combination of "plane wave" functions of z. These plane waves here either have the form $|(y-z)\cdot x|^k$ or $((y-z)\cdot x)^k \log|(y-z)\cdot x|$. A different solution of the same problem is of course given by the Fourier integral representing f:

$$f(z) = \int g(y)e^{iz\cdot y}\,dy,$$

which decomposes $f(z)$ into the plane wave functions $e^{iy\cdot z}$. The advantage of the formulae (1.10), (1.11) is that the integrals contain f itself instead of its Fourier transform.

Formulae (1.10), (1.11) can also be interpreted as expressing $f(z)$ in terms of the integrals of f over hyper-planes. For $|x|=1$

$$J(x, p) = \int\limits_{y\cdot x=p} f(y)dS_y \tag{1.12}$$

represents the integral of f over the hyperplane with unit normal x and (signed) distance p from the origin. By definition (1.12) $J(x, p) = J(-x, -p)$. Taking for an odd n formula (1.10) with $k=1$ we have

$$\begin{aligned}\int\limits_{\Omega_x}\int f(y)\,|(y-z)\cdot x|\,d\omega_x dy &= \int\limits_{\Omega_x} d\omega_x \int\limits_{-\infty}^{+\infty} |p|\,dp \int\limits_{(y-z)\cdot x=p} f(y)\,dS_y \\ &= \int\limits_{\Omega_x} d\omega_x \int\limits_{-\infty}^{+\infty} |p|\,J(x, p+z\cdot x)dp\end{aligned} \tag{1.13}$$

Observing that for $|x|=1$

$$\Delta_z \int\limits_{-\infty}^{+\infty} |p|\,J(x, p+z\cdot x)\,dp$$

$$= \Delta_z\left[\int\limits_{z\cdot x}^{\infty} (p-z\cdot x)J(x, p)dp - \int\limits_{-\infty}^{z\cdot x} (p-z\cdot x)J(x\cdot p)dp\right] = 2J(x, z\cdot x)$$

we find from (1.10) that for odd n

$$2(2\pi^{1}f(z)i)^{n-} = (\Delta_z)^{(n-1)/2} \int\limits_{\Omega_x} J(x,\ x \cdot z)d\omega_x \tag{1.14}$$

Here the integral represents (except for a constant factor ω_n) the average of the plane integrals of f for the planes passing through the point z. A similar formula can be derived for even n from (1.11) with $k = 0$. We notice here that for $|x| = 1$

$$\Delta_z \int\limits_{-\infty}^{+\infty} \log |p|\, J(x, p + z \cdot x)dp = \int\limits_{-\infty}^{+\infty} (\log |p|)J_{pp}(x, p + z \cdot x)dp$$

$$= \int\limits_{-\infty}^{+\infty} (\log |p - z \cdot x|)\, J_{pp}(x, p)dp = -\int\limits_{-\infty}^{+\infty} \frac{1}{p - z \cdot x} J_p(x, p)dp$$

$$= -\int\limits_{p=-\infty}^{p=+\infty} \frac{dJ(x, p)}{p - x \cdot z}$$

where in the last two integrals the Cauchy principal value is to be taken. Then from (1.11) for even n

$$(2\pi i)^n f(z) = (\Delta_z)^{(n-2)/2} \int\limits_{\Omega_x} d\omega_x \int\limits_{p=-\infty}^{p=+\infty} \frac{dJ(x, p)}{p - z \cdot x} \tag{1.15}$$

Expressions equivalent to (1.14), (1.15) for a function $f(z)$ in terms of its plane integrals J were first given by J. Radon [16]. Expressions of a different type, related to the solution of the wave equation have been given by Ph. Mader. [17]

[16] See Radon [1], Bureau [9], ch. IX.
[17] See Mader [1].

CHAPTER II

The Initial Value Problem for Hyperbolic Homogeneous Equations with Constant Coefficients

Hyperbolic equations

The differential equations considered in this chapter shall be of the form

$$(2.1) \qquad L[u] = Q\left(\frac{\partial}{\partial x_1}, \ldots, \frac{\partial}{\partial x_n}, \frac{\partial}{\partial t}\right) u = 0$$

where $Q(\eta_1, \ldots, \eta_n, \lambda)$ is a form of degree m in its arguments with constant real coefficients. The *Cauchy problem* to be solved here consists in finding a solution u of (2.1) satisfying the initial conditions for $t = 0$.

$$(2.2) \qquad \frac{\partial^k u}{\partial t^k} = \begin{cases} 0 & \text{for } k = 0, \ldots, m-2 \\ f(x_1, \ldots, x_n) & \text{for } k = m-1. \end{cases}$$

Once this problem has been solved for arbitrary $f(x)$, it is easy to obtain the solution of the more general Cauchy problem, in which a solution of the differential equation

$$L[u] = w(x, t)$$

with initial data

$$\left(\frac{\partial^k u}{\partial t^k}\right)_{t=0} = f_k(x) \text{ for } k = 0, 1, \ldots, m-1$$

is to be determined.[18] Putting $u = u' + u''$ with

$$u' = \sum_{k=0}^{m-1} \frac{1}{k!} f_k(x) t^k$$

one has to determine u'' as a solution of

$$L[u''] = w - L[u'] = w'(x, t)$$

[18] See Courant-Hilbert [2], vol. II.

with vanishing initial data. By *Duhamel's principle*

$$u''(x, t) = \int_0^t W(x, t - \tau, \tau)d\tau,$$

where $W(x, t, \tau)$ is for each τ a solution of $L[W] = 0$ with initial data

$$Q(0, 1)\left(\frac{\partial^k W(x, t, \tau)}{\partial t^k}\right)_{t=0} = \begin{cases} 0 & \text{for } k = 0, \ldots, m-2 \\ w'(x, \tau) & \text{for } k = m-1. \end{cases}$$

The function $f(x)$ is assumed to vanish outside a bounded set and to be of class C_s, where s is sufficiently large (e.g. $s \geqq 2+m+n$). It can be shown that a solution u of this problem will in general exist only (even for f of class C_∞), if the operator L is *hyperbolic* with respect to the plane $t = 0$. For equations of the form considered here this hyperbolicity is equivalent to the condition that for any real η the equation $Q(\eta, \lambda) = 0$ has only real roots λ, and that the coefficient $Q(0, 1)$ of the highest t-derivative in (2.1) does not vanish.[19] Without restriction of generality we shall assume that $Q(0, 1) = 1$.

Equation (2.1) will be called *strictly hyperbolic* if for any real vector $\eta \neq 0$ all roots λ of

$$Q(\eta, \lambda) = 0 \tag{2.3}$$

are real *and distinct.*

Geometry of the normal surface for a strictly hyperbolic equation [20]

In the case of a strictly hyperbolic equation the *characteristic equation* (2.3) has for real $\eta \neq 0$ exactly m real distinct roots $\lambda_1, \ldots, \lambda_m$. We number them in a unique fashion so that

$$\lambda_1 > \lambda_2 > \ldots > \lambda_m \tag{2.4}$$

[19] This definition would not be sufficient, if Q were not homogeneous. See Gårding [1] for the definition of hyperbolicity for general linear equations with constant coefficients.

[20] See Bureau [1], pp. 165—6; Brusotti [1].

Let η be restricted to the unit sphere Ω_η. Then the λ_k are bounded uniformly, since the coefficient of λ^m in (2.3) has the value 1 and the other coefficients are bounded. Since the totality of roots depends continuously on the coefficients, and since (2.4) holds for all η on Ω_η, it follows that the λ_k are continuous functions $\lambda_k(\eta)$ on the unit sphere.

The equation $Q(\eta, \lambda) = 0$ implies $Q(-\eta, -\lambda) = 0$. It follows that $Q(-\eta, -\lambda_k(\eta)) = 0$. Thus the $-\lambda_k(\eta)$ are the roots belonging to the vector $-\eta$. Since

$$-\lambda_1(\eta) < -\lambda_2(\eta) < \dots < -\lambda_m(\eta)$$

we must have

$$-\lambda_k(\eta) = \lambda_{m-k+1}(-\eta) \text{ for } k = 1, \dots, m \tag{2.5}$$

In the important special case, where

$$Q(\eta, 0) \neq 0 \text{ for all } \eta \text{ on } \Omega_\eta, \tag{2.6}$$

no λ_k can become 0. It follows from the continuity of the $\lambda_k(\eta)$ that the number of positive λ_k is the same for all η. By (2.5) the number of positive $\lambda_k(\eta)$ equals the number of negative $\lambda_k(-\eta)$. Hence under the assumption (2.6) the numbers of positive and negative λ_k are equal for all η. Condition (2.6) can only be satisfied for an *even* order m.

We can interpret $\eta_1, \dots, \eta_n, \lambda$ as homogeneous coordinates of a point in n-dimensional projective space π_n, and $\eta_1/\lambda, \dots, \eta_n/\lambda$ as inhomogeneous coordinates in the euclidean space obtained from π_n by omitting the *plane at infinity* $\lambda = 0$. The origin corresponds to $\eta = 0$. The points of π_n whose homogeneous coordinates satisfy the characteristic equation (2.3) form the *normal surface* Σ. Because of $Q(0, 1) = 1$ the origin is not a point of Σ. For fixed η with $|\eta| = 1$ and varying λ we obtain a line through the origin. Replacing η by $-\eta$ gives the same line. Strict hyperbolicity of the differential equation (with respect to $t = 0$) means that every line through the origin intersects Σ in m real distinct points. The points with homogeneous coordinates $\eta_1, \ldots, \eta_n, \lambda_k(\eta)$ form a connected sheet Σ_k of Σ in π_n. By (2.5) the sheets Σ_k and Σ_{m-k+1} are identical. Thus for even m the normal surface consists of $m/2$ separate sheets, each of which is intersected by a line through the origin in two distinct points. For odd m there are $(m+1)/2$ separate sheets. Of these $(m-1)/2$ are intersected by lines through the origin in two points each. The remaining sheet, $\Sigma_{(m+1)/2}$, is intersected by a line through the origin in only one point, and hence does not divide the space π_n into two parts.

With a sheet Σ_k, that has $k \neq (m+1)/2$, we can associate two point sets R_k and $\overline{R}_k$. Any point P of π_n can be written in the form $(\eta_1, \ldots, \eta_n, \lambda) = (\eta, \lambda)$, where $|\eta| = 1$. Here $\lambda = \infty$ for the origin 0. For $P \neq 0$ there are exactly two such representations (η, λ) and $(-\eta, -\lambda)$. The set R_k is defined as consisting of the points (η, λ) with

$$0 < \frac{\lambda - \lambda_k(\eta)}{\lambda - \lambda_{m-k+1}(\eta)} < \infty,$$

whereas for points in $\overline{R}_k$

$$-\infty < \frac{\lambda - \lambda_k(\eta)}{\lambda - \lambda_{m-k+1}(\eta)} < 0.$$

The definition of R_k and $\overline{R}_k$ does not depend on which of the two representations is chosen for P, by virtue of (2.5). The sets Σ_k, R_k, $\overline{R}_k$ fill the whole space π_n. The sheet Σ_k is the boundary for both R_k and $\overline{R}_k$, and divides π_n into those two sets. Every set R_k contains the origin 0. Every point P of R_k can be joined in R_k to 0 by a projective straight line "segment" (i.e. a sub set of a projective line bounded by two points). It follows that R_k is homeomorphic to the interior of a sphere. By (2.4) each set R_k with $k < (m+1)/2$ contains all sets with smaller index.

Every line through a point P of R_k intersects Σ_k in at least two distinct real points, if $k < (m+1)/2$. For let P be a point of R_k and l be a line through P that has one or no point in common with Σ_k. Let π_2 be the 2-dimensional projective plane through

0 and l. Then π_2 intersects Σ_k in a curve σ_k. The line l has at most one point in common with σ_k, while every line in π_2, which passes through 0 has two distinct points in common with σ_k. We can introduce l as line at infinity of π^2. One of the two rays pointing from 0 to P will be contained in R_k and hence will be free of points of σ_k. We rotate this ray in either direction about 0. Rotating in one or in the other direction one must arrive at a first ray, which contains a point of σ_k (possibly at infinity) since every line through 0 does contain points of σ_k. These first rays containing points of σ_k cannot have them at a finite distance, because in that case they would have to have a multiple intersection with σ_k, which is excluded. Thus they intersect σ_k at infinity. Since the line at infinity contains only one point of σ_k by assumption, the two first rays from 0 with points of σ_k must be diametrically opposite. Then however the line formed by the two rays, which also passes through 0, would have only one point of intersection in common with σ_k, which leads to a contradiction.

A line through a point of R_1 will then intersect each sheet Σ_k with $k < (m + 1)/2$ in exactly two points, and the sheet $\Sigma_{(m+1)/2}$ in one point, since there are only m intersections available and imaginary intersections occur in pairs. Let P_1 and P_2 be points of R_1. Then the line joining P_1 and P_2 intersects Σ_1 in exactly two points, which cannot separate P_1 and P_2, since Σ_1 divides space into two parts. It follows that R_1 is convex in the sense that with every two points one of the two projective line "segments" joining the points belongs to R_1.

The stucture of the normal surface considered as a surface in *euclidean* space can be more complicated. Each of the connected components Σ_k of the normal surface can fall apart into several sheets, if the plane at infinity is omitted. Only in the case of even n with condition (2.6) satisfied will the normal surface be a bounded surface in euclidean space consisting of $m/2$ ovals containing one another, with the innermost oval *convex* in the ordinary sense.

Solution of the Cauchy problem for a strictly hyperbolic equation

The equations of the form (2.1), in which all terms are of the same order, admit plane wave solutions of arbitrary "shape." Let $g(s)$ be an arbitrary function of class C_m. Then

$$u = g((x - y) \cdot \eta + t\lambda) \tag{2.7}$$

will be a solution of (2.1) of class C_m, if the constants η and λ satisfy the characteristic equation (2.3). We shall solve the Cauchy problem (2.1), (2.2) first for the case, where $f(x)$ is of the form $|(y - x) \cdot \eta|^k$ or $[(y - x) \cdot \eta]^k \log |(y - x) \cdot \eta|$. The solution u for that case is a simple combination of plane wave solutions of the type (2.7). The solution for general f follows then by superposition from the formulae (1.10), (1.11).

Let for $|\eta| = 1$ the roots of (2.3) be given by $\lambda_k = \lambda_k(\eta)$. We shall make frequent use of the fact that for a polynomial $Q(\lambda)$ of degree m with simple roots λ_k and highest coefficient 1 the identity

$$\frac{1}{2\pi i} \oint_C \frac{\lambda^q}{Q(\lambda)} d\lambda = \sum_{k=1}^{m} \frac{\lambda_k^q}{Q_\lambda(\lambda_k)} = \begin{cases} 0 & \text{for } q = 0, 1, \ldots, m-2 \\ 1 & \text{for } q = m - 1 \end{cases} \tag{2.8}$$

holds, where C is a path in the complex λ-plane enclosing all λ_k. It follows that for $g(s)$ of class C_m

$$u = \sum_{k=1}^{m} \frac{g((x - y) \cdot \eta + \lambda_k t)}{Q_\lambda(\eta, \lambda_k)} \tag{2.9}$$

is a solution of (2.1) with initial data for $t = 0$

$$\frac{\partial^r u}{\partial t^r} = \begin{cases} 0 & \text{for } r = 0, 1, \ldots, m - 2 \\ g^{(m-1)}((x - y) \cdot \eta) & \text{for } r = m - 1 \end{cases} \tag{2.10}$$

Here use has been made of the assumption that the λ_k are real and distinct. Since the $\lambda_k(\eta)$ are continuous functions of n on the unit sphere, we have then in

$$Z(x - y, t) = \int_{\Omega_\eta} \sum_{k=1}^{m} \frac{g((x - y) \cdot \eta + t\lambda_k(\eta))}{Q_\lambda(\eta, \lambda_k(\eta))} d\omega_\eta \tag{2.11}$$

a solution of (2.1), for wich

$$\left(\frac{\partial^r Z}{\partial t^r}\right)_{t=0} = \begin{cases} 0 & \text{for } r = 0, 1, \ldots, m-2 \\ \displaystyle\int_{\Omega_\eta} g^{(m-1)}((x-y)\cdot\eta)\,d\omega_\eta & \text{for } r = m-1. \end{cases} \tag{2.12}$$

If $f(y)$ is an arbitrary continuous function vanishing outside a bounded set, the function

$$v(x, t) = \int f(y) Z(x-y, t)\,dy \tag{2.13}$$

will again be a solution of $L[v] = 0$ of class C_m. Initially v and its derivatives of order $\leqq m-2$ will vanish, and

$$\frac{\partial^{m-1} v}{\partial t^{m-1}} = \int_{\Omega_\eta}\int g^{(m-1)}((x-y)\cdot\eta)\, f(y)\,dy\,d\omega_\eta. \tag{2.14}$$

We choose now for $g(s)$ the function

$$g(s) = \frac{s^{m-1+q}\operatorname{sign} s}{4(m-1+q)!\,(2\pi i)^{n-1}} \quad \text{for odd } n \tag{2.15a}$$

and

$$g(s) = \frac{s^{m-1+q}\log|s|}{-(m-1+q)!\,(2\pi i)^{n}} \quad \text{for even } n. \tag{2.15b}$$

Here q shall be an integer with

$$q \geqq 2, \quad q + n \text{ even.}$$

Then

$$g^{(m-1)}(s) = \begin{cases} \dfrac{|s|^q}{4q!\,(2\pi i)^{n-1}} & \text{for odd } n \\[2ex] \dfrac{-s^q(\log|s| + c)}{q!\,(2\pi i)^n} & \text{for even } n \end{cases} \tag{2.15c}$$

with a certain constant c. It follows from formulae (1.10), (1.11) that for f in C_1

$$(2.16) \qquad (\Delta_x)^{(n+q)/2}\left[\frac{\partial^{m-1}v(x,t)}{\partial t^{m-1}}\right]_{t=0} = f(x).$$

(The constant c contributes only a polynomial in x of degree q to the right hand side of (2.14), which is annihilated by the Δ-operators).

Assume now that $f(x)$ is actually of class C_{n+q}. We can write v in the form

$$v(x,t) = \int f(y+x)\,Z(-y,t)\,dy$$

and form

$$(2.17) \quad u(x,t) = (\Delta_x)^{(n+q)/2}v(x,t) = \int [(\Delta_y)^{(n+q)/2}f(y)]\,Z(x-y,t)\,dy.$$

Since $(\Delta_y)^{(n+q)2}f(y)$ is continuous, u is a solution of (2.1) of class C_m. Moreover initially u and its derivatives of order $\leq m-2$ vanish. We have

$$\frac{\partial^{m-1}u}{\partial t^{m-1}} = \int [(\Delta_y)^{(n+q)/2}f(x+y)]\frac{\partial^{m-1}Z(-y,t)}{\partial t^{m-1}}\,dy$$

$$= (\Delta_x)^{(n+q)/2}\int f(x+y)\frac{\partial^{m-1}Z(-y,t)}{\partial t^{m-1}}\,dy$$

$$= (\Delta_x)^{(n+q)/2}\frac{\partial^{m-1}}{\partial t^{m-1}}v(x,t).$$

For $t=0$ it follows then from (2.16) that

$$\frac{\partial^{m-1}u}{\partial t^{m-1}} = f(x).$$

Hence $u(x,t)$ is a solution of the initial value problem (2.1), (2.2). For q we can choose the number 3, if n is odd, and the number 2, if n is even. Thus it is sufficient to assume f to be of class C_{n+3} for odd n and to be of class C_{n+2} for even n.

In the case where $n < m-1$ we can perform the differentiations occurring in the expression for u under the integral sign. We have from (2.15 a, b)

$$g^{(q+n)}(s) = \begin{cases} \dfrac{s^{m-1-n}\operatorname{sign} s}{4(m-n-1)!(2\pi i)^{n-1}} & \text{for odd } n \\[2ex] \dfrac{s^{m-n-1}(\log|s|+c)}{-(m-n-1)!(2\pi i)^{n}} & \text{for even } n \end{cases}$$

with a certain constant c. Putting

$$K(x-y,\ t) = (\Delta_x)^{(n+q)/2} Z(x-y,\ t) \tag{2.18}$$

$$F_k = (x-y)\cdot\eta + t\lambda_k(\eta), \text{ for } k = 1,\ldots, m \tag{2.19}$$

we have from (2.11)

$$(2\pi i)^n(m-n-1)!K(x-y,t) = \begin{cases} \dfrac{\pi i}{2}\displaystyle\sum_{k=1}^{m}\int_{\Omega_\eta} \frac{F_k^{m-n-1}\operatorname{sign} F_k}{Q_\lambda(\eta, \lambda_k(\eta))} d\omega_\eta \\[3ex] -\displaystyle\sum_{k=1}^{m}\int_{\Omega_\eta} \frac{F_k^{m-n-1}\log|F_k|}{Q_\lambda(\eta, \lambda_k(\eta))} d\omega_\eta \end{cases} \tag{2.20}$$

respectively for odd and even n. (The constant c makes no contribution by (2.8), since F_k^{m-n-1} is a polynomial of degree $< m-1$ in λ_k). Then from (2.17), (2.13)

$$u(x,t) = \int f(y)K(x-y,\ t)dy. \tag{2.21}$$

This expression for u could in general also be used in the case where $n = m-1$, but would require some justification, since in that case the integrands in (2.20) become singular. (See p. 29 below).

Expression of the kernel by an integral over the normal surface

We can transform the expression (2.20) for K into an integral over the normal surface. For this purpose we make the further assumption that L does not contain $\partial/\partial t$ as a factor. This amounts to the statement that none of the functions $\lambda_k(\eta)$ vanishes identically or that the plane at infinity does not form one of the sheets of the normal surface. In this case the set of points on the unit sphere Ω_η for which $Q(\eta, 0) = 0$ forms a lower dimensional manifold and hence the contribution of the points η with $\lambda_k(\eta)=0$ can be neglected in forming the integrals (2.20) for K.

For η varying over the points of Ω_η with $\lambda_k(\eta) \neq 0$ the point

$$\xi = \eta/\lambda_k(\eta)$$

varies over the bounded portion of the sheet Σ_k of the normal surface $Q(\xi, 1) = 0$ referred to inhomogeneous coordinates. The solid angle $d\omega_\eta$ is related to the corresponding element of surface dS of Σ_k by the equation (see Fig. 2)

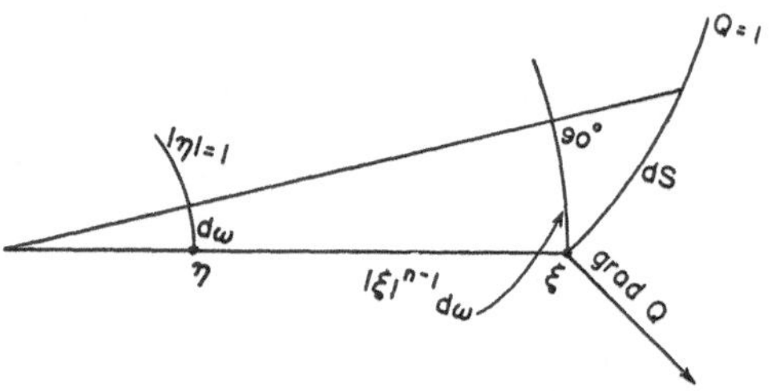

Figure 2

$$dS = \frac{|\xi|^n \,|\text{grad}\, Q(\xi, 1)|}{|\sum_i Q_{\xi_i}(\xi, 1)\xi_i|}\, d\omega_\eta, \tag{2.21a}$$

where $|\text{grad}\, Q(\xi, 1)| = \sqrt{\sum_i Q^2_{\xi_i}(\xi, 1)}$. Here $|\xi| = 1/\,|\lambda_k(\eta)|$ and

$$\Big|\sum_i Q_{\xi_i}(\xi, 1)\xi_i\Big| = |-Q_\lambda(\xi, 1)| = |\xi|^{m-1}\,|Q_\lambda(\eta, \lambda_k(\eta))|.$$

Put

$$E = F_k/\lambda_k(\eta) = (x-y)\cdot\xi + t.$$

Then formula (2.20) yields for odd n

$$4(2\pi i)^{n-1}(m-n-1)!\,K(x-y, t)$$
$$= \sum_k \int_{\Sigma_k} \frac{E^{m-n-1}\,(\text{sign}\,\lambda_k)^{m-1}\,\text{sign}\,E}{|\text{grad}\,Q(\xi, 1)|\,\text{sign}\,Q_\lambda(\eta, \lambda_k)}\, dS$$

Since $Q(0, 1) = 1$ the function $Q(\eta, \lambda)$ is positive for large positive λ. It follows from (2.4) that

$$\operatorname{sign} Q_\lambda(\eta, \lambda_k) = (-1)^{k+1}.$$

Each point ξ of the surface $Q(\xi, 1) = 0$ makes two contributions to the expression for K, since ξ must be representable in the forms

$$\xi = \eta/\lambda_k(\eta) = -\eta/\lambda_{m+1-k}(-\eta)$$

with a certain k. Here $\lambda_k(\eta)$ and $\lambda_{m+1-k}(-\eta)$ have opposite signs. Hence

$$(-1)^k(\operatorname{sign} \lambda_k(\eta))^{m-1} = (-1)^{m-k+1}(\operatorname{sign} \lambda_{m+1-k}(-\eta))^{m-1}.$$

It follows that both representations of ξ make the same contribution. It is then sufficient to take twice the contribution arising from the representation of ξ with a positive λ_k. For positive λ_k the number $k-1$ equals the number of positive roots of the equation $Q(\eta, \lambda) = 0$, which exceed λ_k or also the number of points of the normal surface which lie on the same ray from the origin and are closer to the origin. Consequently for odd $n < m - 1$

$$(2.22) \qquad 2(2\pi i)^{n-1}(m-n-1)!K(x-y, t) = \int_{Q(\xi,1)=0} \frac{(-1)^N E^{m-n-1} \operatorname{sign} E}{|\operatorname{grad} Q(\xi, 1)|} dS$$

where $N = N(\xi)$ is the number of points of the normal surface on the same ray as ξ and closer to the origin. It is clear that $(-1)^N$ is constant along each connected component of the normal surface in euclidean space, but may be different for the different components making up the same sheet Σ_k of the normal surface in projective space.

A similiar formula can be derived from (2.20) for even n. It is convenient in that case to replace the term $\log |F_k|$ in (2.20) by $\log |F_k/F_0|$, where $F_0 = (x-y) \cdot \eta$. The term $\log |F_0|$, being independent of k, makes no contribution to the sum over k

by virtue of the identities (2.8). Putting

$$E_0 = (x - y) \cdot \xi$$

we obtain for even $n < m - 1$

$$(2.23) \qquad (2\pi i)^n (m - n - 1)! K(x - y, t) = -2 \int\limits_{Q(\xi, 1) = 0} \frac{(-1)^N E^{m-n-1} \log | E/E_0 |}{| \operatorname{grad} Q(\xi, 1) |} dS.$$

Formulae (2.22), (2.23) were given by Herglotz for $n \leqq m - 1$ under the assumption that the normal surface is bounded (and hence m is even).[21] If we introduce $\xi_1, \ldots, \xi_{n-1}$ as variables of integration we have

$$\frac{dS}{| \operatorname{grad} Q(\xi, 1) |} = \pm \frac{d\xi_1 \ldots d\xi_{n-1}}{Q_{\xi_n}(\xi, 1)}$$

Splitting up the domain of integration into portions with $E > 0$ and $E < 0$ we can let the variables of integration assume complex values and obtain expressions of the type given by F. Bureau.[22] An explicit evaluation of the multiple integrals for K or even a reduction to single integrals is, of course, not feasible in general. In special cases, particularly when Q can be factored into quadratic factors, the integrals have been evaluated and discussed by Bureau.[23]

We now turn to the case, where $n \geqq m - 1$. In that case only a certain number of the Laplaceans appearing in the expression (2.17) can be carried out under the integral sign and applied to Z. It is convenient here to replace the Δ-operators occurring by t-derivatives. For that purpose however we have to restrict ourselves now to the case, where $Q(\eta, 0) \neq 0$ on Ω_η and hence where m is even. With this assumption the normal surface is bounded, and $\lambda_k(\eta) \neq 0$ for all η on Ω_η.

We use for f and g of class C_2 the identity

[21] See Herglotz [2], p. 294. A solution for odd m is given by Petrovskii [1].
[22] See Bureau [2], [3], [4].
[23] See Bureau [5], [6], [7].

$$
\begin{aligned}
\frac{\partial^2}{\partial t^2} \int f(y) \Bigl(\int_{\Omega_\eta} \frac{g(F_k)\lambda_k^{-\alpha}}{Q_\lambda(\eta, \lambda_k(\eta))} d\omega_\eta \Bigr) dy \\
&= \int f(y) \Bigl(\int_{\Omega_\eta} \frac{g''(F_k)\,\lambda_k^{2-\alpha}}{Q_\lambda(\eta, \lambda_k(\eta))} d\omega_\eta \Bigr) dy \\
&= \int f(y) \Delta_y \Bigl(\int_{\Omega_\eta} \frac{g(F_k)\,\lambda_k^{2-\alpha}}{Q_\lambda(\eta, \lambda_k(\eta))} d\omega_\eta \Bigr) dy \\
&= \int (\Delta_y f(y)) \Bigl(\int_{\Omega_\eta} \frac{g(F_k)\,\lambda_k^{2-\alpha}}{Q_\lambda(\eta, \lambda_k(\eta))} d\omega_\eta \Bigr) dy.
\end{aligned}
$$

Repeated use of this identity, assuming that f is of class C_{n+q} and vanishing outside a bounded set, transforms the expression (2.17) for u into

$$
u(x, t) = \frac{\partial^{n+q}}{\partial t^{n+q}} \int f(y)\, Z'(x - y, t)\, dy
$$

where

$$
Z'(x - y, t) = \sum_{k=1}^{m} \int_{\Omega_\eta} \frac{g(F_k)\,\lambda_k^{-n-q}}{Q_\lambda(\eta, \lambda_k(\eta))} d\omega_\eta
$$

(Here use is made of $\lambda_k \neq 0$.) Now

$$
F_k = (x - y) \cdot \eta + t\lambda_k(\eta)
$$

cannot vanish identically in η, for in that case $Q(\eta, \lambda)$ would have to contain the linear function $(x - y) \cdot \eta + t\lambda$ as a factor, the normal surface would contain a plane, and hence would extend to infinity, contrary to assumption. Thus the set of points on Ω_η with $F_k = 0$ forms a lower dimensional point set. It follows that $Z'(x - y, t)$ with g given by (2.15a, b) has continuous t-derivatives of orders $\leq m - 1 + q$. Then

$$
u(x, t) = \frac{\partial^{n-m+1}}{\partial t^{n-m+1}} \int f(y) K'(x - y, t)\, dy \tag{2.24}
$$

where

$$K'(x-y,t) = \frac{\partial^{m-1+q}}{\partial t^{m-1+q}} Z'(x-y,\ t)$$

$$= \Sigma_k \int\limits_{\Omega_\eta} \frac{\lambda_k^{m-1-n} g^{(m-1+q)}(F_k)}{Q_\lambda(\eta,\lambda_k(\eta))} d\omega_\eta .$$

Here

$$g^{(m-1+q)}(s) = \begin{cases} \dfrac{\operatorname{sign} s}{4(2\pi i)^{n-1}} & \text{for odd } n \\[2ex] \dfrac{\log|s|+c}{-(2\pi i)^n} & \text{for even } n. \end{cases}$$

with a certain constant c. The constant c makes only a constant contribution to K', which would be annihilated by the t-derivatives occurring in the expression for u. (For even n and m the value of $n-m+1$ is at least 1). For the same reason we can replace $\log|F_k|$ by $\log|F_k/F_0|$, since the contribution of $\log|F_0|$ does not depend on t. We are thus led to the expressions

$$K'(x-y,t) = \frac{1}{4(2\pi i)^{n-1}} \sum_k \int\limits_{\Omega_\eta} \frac{\lambda_k^{m-1-n}\operatorname{sign} F_k}{Q_\lambda(\eta,\lambda_k(\eta))} d\omega_\eta$$

for odd n, and

$$K'(x-y,\ t) = \frac{-1}{(2\pi i)^n} \sum_k \int\limits_{\Omega_\eta} \frac{\lambda_k^{m-1-n}\log|F_k/F_0|}{Q_\lambda(\eta,\lambda_k(\eta))} d\omega_\eta .$$

Transforming to the normal surface as before we obtain finally

$$(2.25)\quad K'(x-y,t) = \frac{1}{2(2\pi i)^{n-1}} \int\limits_{Q(\xi,1)=0} \frac{(-1)^N \operatorname{sign} E}{|\operatorname{grad} Q(\xi,1)|} dS$$

for odd n, and

$$(2.26)\quad K'(x-y,t) = \frac{-2}{(2\pi i)^n} \int\limits_{Q(\xi,1)=0} \frac{(-1)^N \log|E/E_0|}{|\operatorname{grad} Q(\xi,1)|} dS$$

for even n. For $n = m-1$ and a bounded normal surface these formulae agree with the expressions that would be obtained from (2.22) and (2.23).

The domain of dependence

Let x, y, t be such that the plane $E = (x - y) \cdot \xi + t$ does not intersect the normal surface. Then for odd n the sign of E is constant on the surface, and hence K' is constant by (2.25). It follows from (2.24) that for odd $n > m - 1$ the solution does not depend on those initial values $f(y)$ corresponding to y for which $-(x - y)/t$ lies in the polar reciprocal region of the convex hull of the sheet $\Sigma_{m/2}$ of the normal surface with respect to the sphere $|\xi|^2 - 1 = 0$ of radius 1. Thus a given point y is not included in the domain of dependence of the solution $u(x, t)$ for given x, if t is sufficiently large. We thus have a "lacuna" in the domain of dependence, [24] and in a certain sense a generalization of Huygens' principle in its strong form to homogeneous equations of higher order in an odd number of dimensions. For $n \leqq m - 1$ we can conclude from (2.22) only that K will be a form of degree $\leqq m - n - 1$ in the corresponding region.

Following Herglotz [25] we can prove in the case of odd $n \leqq m - 1$ that *the domain of dependence of u on the initial values is bounded.* For this purpose we observe that by (2.20), (2.8)

$$K(x - y,\ 0) = 0,$$

since for $t = 0$ the expression F_k does not depend on k. It follows from (2.22) that

$$(2.27)\qquad \int\limits_{\Sigma} \frac{(-1)^N [(x-y)\cdot\xi]^{m-n-1}\,\mathrm{sign}\,[(x-y)\cdot\xi]}{|\mathrm{grad}\,Q(\xi)|}\,dS = 0,$$

if Σ is an algebraic surface of degree m and equation $Q(\xi) = 0$, which is such that every line through the origin intersects it

[24] See Petrovskii [1] Gårding [1], p. 60. The lacuna is the same if the special initial data (2.2) are replaced by more general ones.

[25] See Herglotz [1], part II, p. 295, Gårding [1], where the proof is given for even n as well. The same result can be proved much more generally for linear equations with analytic coefficients by the method of Holmgren. See John [3], Myskis [1], Courant-Hilbert [2].

in m distinct real points in projective space and such that the plane at infinity is not a sheet of Σ. For more general t the right hand side of (2.22) can be brought into the form (2.27) by the substitution

$$\xi = \xi_0 + \xi',$$

where ξ_0 is chosen in such a way that

$$(x - y) \cdot \xi_0 + t = 0.$$

Put $Q(\xi_0 + \xi', 1) = Q'(\xi')$. Then the right hand side of (2.22) becomes

$$(2.28) \qquad \int_{\Sigma'} \frac{(-1)^N [(x-y) \cdot \xi']^{m-n-1} \operatorname{sign} [(x-y) \cdot \xi']}{| \operatorname{grad} Q'(\xi') |} dS$$

where Σ' is the m-th degree surface $Q'(\xi') = 0$. If here ξ_0 is a point of the projectively convex set R_1 bounded by Σ_1, every line through the point $\xi' = 0$ will intersect Σ' in m distinct real points, as shown on p. 19. It remains to show that $(-1)^N$ has the same or opposite value counted from the point $\xi = 0$ or the point $\xi' = 0$. Let ξ be a point of the sheet Σ_k, where $1 \leqq k \leqq (m+1)/2$. (See Fig. 3.) One of the projective line segments joining 0 and ξ_0 is contained in R_1 and hence is free of points of Σ. Let ζ denote a variable point of that segment. If the segment is the *bounded* one then the number of intersections of the bounded open segment $\zeta\xi$ with Σ does not change as ζ varies from 0 to ξ_0 since there are no multiple intersections with Σ on the segment $\xi\zeta$ and no intersections can lie on the segment $0\xi_0$. In that case then the open bounded segments 0ξ and $\xi_0\xi$ intersect Σ in the same number of points. If the *infinite* line segment $0\xi_0$ is free of points of Σ we let ζ vary from 0 to ξ_0 along that segment. The segment $\xi\zeta$ varies then continuously in projective space starting from the bounded segment $\xi 0$ and ending up with the unbounded segment $\xi\xi_0$. The number of points of Σ on $\xi\zeta$ stays constant. It follows that in this case the number of points of Σ on the open bounded segments 0ξ and $\xi_0\xi$ adds up to $m - 1$, independently of the choice of the point ξ on Σ. We see then that $(-1)^N$ has either the same value throughout the surface when counted from 0 or from ξ_0 or the opposite value. Hence the integral (2.28) vanishes and $K(x-y, t) = 0$ whenever there exists a ξ_0 in R_1 such that

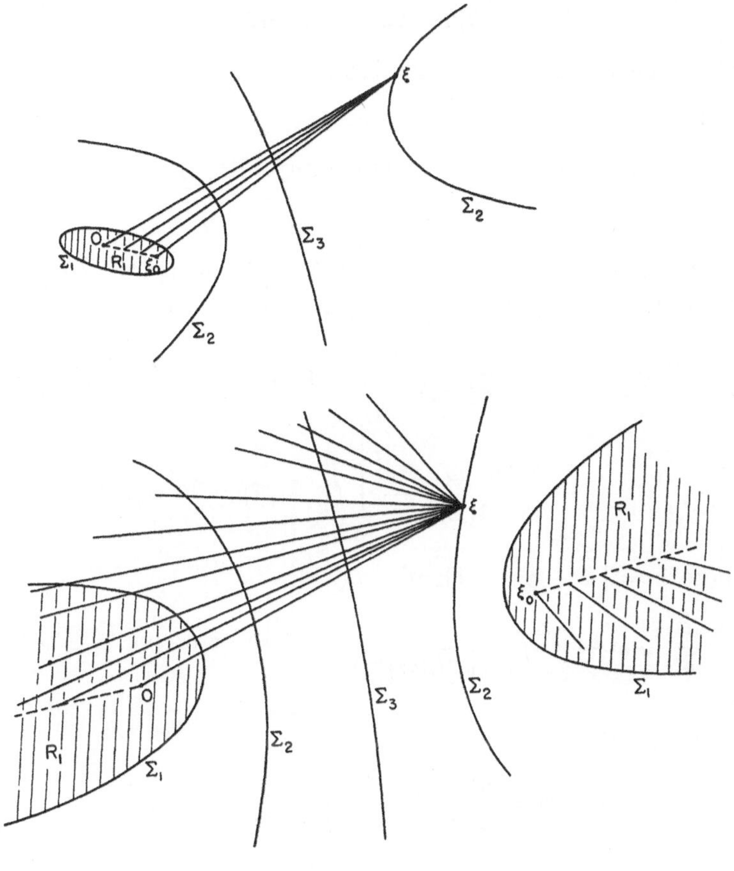

Figure 3

$(x-y)\cdot\xi_0 + t = 0$, i.e. whenever the plane $(x-y)\cdot\xi + t = 0$ intersects R_1. This is certainly the case for sufficiently large $|x-y|$, since the distance of that plane from the interior point 0 of R_1 is given by $t/|x-y|$. The polar reciprocal of R_1 with respect to the unit sphere is a bounded convex set. The solution $u(x, t)$ of the initial value problem does not depend on the values of $f(y)$ in those points y for which $(y - x)/t$ lies outside that convex set.

The wave equation

For the equation of second order

$$\Delta_x u = u_{tt} \tag{2.29}$$

we have a normal surface coinciding with the unit sphere Ω_ξ:

$$Q(\xi, 1) = 1 - |\xi|^2 = 0.$$

On the single sheet of this surface $N = 0$ and $|\operatorname{grad} Q(\xi, 1)| = 2$. It follows from (2.25) and (1.2) for odd n that

$$\begin{aligned} K'(x-y, t) &= \frac{1}{4(2\pi i)^{n-1}} \int_{\Omega_\xi} \operatorname{sign}\,[(x-y)\cdot\xi + t]\, d\omega_\xi \\ &= \frac{\omega_{n-1}}{4(2\pi i)^{n-1}} \int_{-1}^{+1} (1-p^2)^{(n-3)/2} \operatorname{sign}\,(rp + t)\, dp \end{aligned}$$

where $r = |x - y|$. Then

$$\frac{\partial}{\partial t} K'(x-y, t) = \begin{cases} \dfrac{\omega_{n-1}}{2(2\pi i)^{n-1}}\, r^{2-n}(r^2 - t^2)^{(n-3)/2} & \text{for } r > t \\ 0 & \text{for } r < t. \end{cases}$$

From (2.24)

$$\begin{aligned} u(x, t) &= \frac{\partial^{n-2}}{\partial t^{n-2}} \int f(y) \frac{\partial}{\partial t} K'(x-y, t)\, dy \\ &= \frac{\partial^{n-2}}{\partial t^{n-2}} \frac{\omega_n \omega_{n-1}}{2(2\pi i)^{n-1}} \int_t^\infty I(x, r) r (r^2 - t^2)^{(n-3)/2}\, dr, \end{aligned}$$

where $I(x, r)$ is the spherical mean of f:

$$I(x, r) = \frac{1}{\omega_n} \int_{|x-y|=r} f(y) r^{1-n}\, dS_y \tag{2.30}$$

Here from (1.4)

$$\omega_n \omega_{n-1} = \frac{2^n \pi^{n-1}}{(n-2)!} \tag{2.31}$$

If we use the fact that

$$\int_0^\infty I(x, r) r (r^2 - t^2)^{(n-3)/2} \, dr$$

is a polynomial in t of degree $n-3$, whose $(n-2)$nd t-derivative vanishes, we obtain the classical formula [26]

$$u(x, t) = \frac{1}{(n-2)!} \frac{\partial^{n-2}}{\partial t^{n-2}} \int_0^t I(x, r) r (t^2 - r^2)^{(n-3)/2} \, dr. \tag{2.32}$$

To obtain the same formula for *even* n from (2.26) is more difficult. We have from (2.26) and (1.2)

$$K'(x-y, t) = \frac{-\omega_{n-1}}{(2\pi i)^n} \int_{-1}^{+1} (1-p^2)^{(n-3)/2} \left[\log \left| p + \frac{t}{r} \right| - \log |p| \right] dp$$

Here the term $\log |p|$ can be omitted, since it would make no contribution to $\partial K'/\partial t$. We can write K' as an integral in the complex p-plane taken over a closed path C

$$K'(x-y, t) = \frac{+\omega_{n-1}}{2(2\pi i)^n} \mathcal{R}e \oint_C (1-p^2)^{(n-3)/2} \log \left(p + \frac{t}{r} \right) dp \tag{2.33}$$

The function $(1-p^2)^{(n-3)/2}$ is defined here so as to be univalued outside a slit from -1 to $+1$ and to have positive real values on the upper border of the slit. For $\log s$ the principal branch is taken, defined in the s-plane slit along the positive imaginary axis. Putting $p_0 = -t/r$ we take for C a closed path through the point p_0, which crosses neither the slit from -1 to $+1$ nor

[26] Tedone [1], Courant-Hilbert [1], vol. 2, p. 399. See also Bureau [8], M. Riesz [1], Cooper [1].

the slit from p_0 to $p_0 + i\infty$ in the p-plane. (See Fig. 4.)

The integral in (2.33) is of the form

$$h(p_0) = \oint_C g(p) \log (p - p_0)\, dp, \tag{2.34}$$

where the function $g(p)$ is regular analytic outside a finite slit along the real axis and is continuous from either side in the points

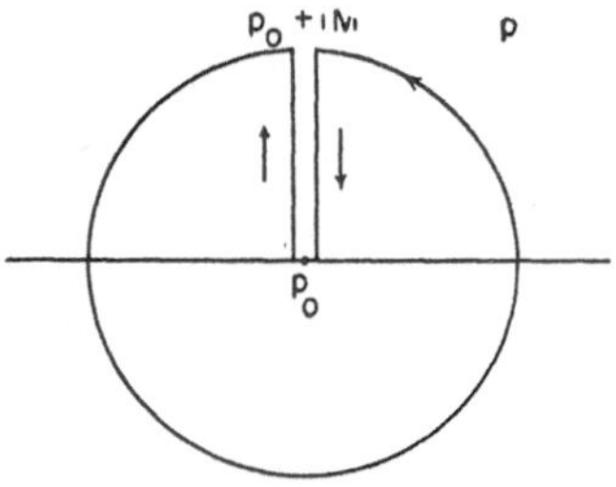

Figure 4

of the slit. (We have to assume here that $n \geqq 4$). We can use as path of integration a circle of radius M about the point p_0 together with a slit from p_0 to $p_0 + iM$. Using the fact that $\log p$ changes by the amount $2\pi i$ on crossing the imaginary axis, we arrive at the expression

$$h(p_0) = \int_{|p|=M} g(p + p_0) \log p\, dp - 2\pi i \int_{p_0}^{p_0+iM} g(p) dp$$

Then

$$h'(p_0) = \int_{|p|=M} g'(p+p_0) \log p\, dp - 2\pi i (g(p_0 + iM) - g(p_0)).$$

Integration by parts applied to the integral yields then the formula

$$h'(p_0) = -\oint_{|p|=M} g(p + p_0) \frac{dp}{p} + 2\pi i\, g(p_0) \tag{2.35}$$

taking into account the multiple valuedness of $\log p$ on the circle. [The value of $g(p_0)$ occurring in (2.35) is more precisely $g(p_0 + i0)$.]

In the special case under consideration

$$g(p) = (1 - p^2)^{(n-3)/2};$$

we get

$$h'(p_0) = -\int_{|p|=M} (1 - (p + p_0)^2)^{(n-3)/2} \frac{dp}{p} + 2\pi i (1 - p_0^2)^{(n-3)/2}.$$

The integral in this formula is the residue of the function $(1/p)[1 - (p + p_0)^2]^{(n-3)/2}$ at $p = \infty$. This is easily seen to be a polynomial $P(p_0)$ of degree $\leqq n - 3$. Then

$$\begin{aligned}\frac{\partial}{\partial t} K'(x - y, t) &= -\frac{1}{r}\frac{\partial K'}{\partial p_0} \\ &= \frac{-\omega_{n-1}}{2r(2\pi i)^n} \mathcal{R}e \left[2\pi i \left(1 - \frac{t^2}{r^2}\right)^{(n-3)/2} + P\left(-\frac{t}{r}\right)\right]\end{aligned}$$

The polynomial P makes no contribution to

$$u(x, t) = \frac{\partial^{n-2}}{\partial t^{n-2}} \int f(y) \frac{\partial}{\partial t} K'(x - y, t) dy$$

The remaining portion of $\partial K'(x - y, t)/\partial t$ has the value 0 for $r > t$ and the value

$$\frac{\omega_{n-1}}{2(2\pi)^{n-1}} r^{2-n} (t^2 - r^2)^{(n-3)/2}$$

for $r < t$ (observe that then $p_0 < -1$). The same argument as before establishes then formula (2.32) for even $n \geqq 4$. Similar proofs show the validity of the formula for $n = 2$ as well.

Formula (2.32) has been established under the assumption that $f(x)$ possesses derivatives of sufficiently high order. For *odd* n we can carry out $(n - 3)/2$ of the t-differentiations under the integral sign without any contributions from the limits of integration. The remaining $(n - 1)/2$ differentiations produce then an expression of the form

$$(2.36) \qquad u(x, t) = \sum_{k=0}^{(n-3)/2} c_k t^{k+1} \frac{\partial^k I(x, t)}{\partial t^k}$$

with certain numerical constants c_k. Now $I(x, t)$ is at least of the same class as $f(x)$. It follows that the u represented by (2.36) is of class C_2 for $f(x)$ of class $C_{(n+1)/2}$. Since the expression (2.36) represents a solution of the initial value problem of the wave equation for all sufficiently regular f, we can conclude, by approximation, that it solves the problem for all f in $C_{(n+1)/2}$. [This is the problem with the special initial values $u = 0$, $u_t = f$ for $t = 0$. For the more general initial values $u = g$, $u_t = f$ the function g would have to be assumed to be of class $C_{(n+3)/2}$].[27] The assumption that f is of class $C_{(n+1)/2}$ cannot be weakened in general, since $I(x, t)$ can be of the same class as $f(x)$. Thus in the case where $f(x)$ is a function with spherical symmetry, $f(x) = h(|x|)$, we have $I(0, t) = h(t)$, and the number of t-derivatives of I is equal to the class of f. For *even* n we can carry out $(n-2)/2$ differentiations under the integral sign in formula (2.32) and find that u represents a solution of the wave equation for f of class $C_{(n+2)/2}$.

The initial value problem for hyperbolic equations with a normal surface having multiple points

So far the assumption that the characteristic equation $Q(\eta, \lambda) = 0$ has no multiple roots had been made. We now abandon this restriction, modifying the expression for $Z(x-y, t)$ used previously in such a way that it has sense in the case of multiple roots. We proceed as in the case of the wave equation with even n. We introduce the function

$$(2.37) \qquad g(s) = \frac{s^{m-1+q}\left(\frac{\pi i}{2} + \log s\right)}{-(m-1+q)!\,(-2\pi i)^n}$$

where log s denotes the principal branch of the logartihm defined

[27] See Diaz and Weinberger [1], p. 709, Cooper [1].

in the plane slit along the positive imaginary axis. Then $\mathscr{R}e\,[g(s)]$ agrees for real s with the functions called g in formulae (2.15a, b). Put

$$F = (x - y) \cdot \eta + \lambda t, \quad F_0 = (x - y) \cdot \eta. \tag{2.38}$$

The expression (2.9) will be replaced by

$$R = \mathscr{R}e \left[\frac{1}{2\pi i} \oint_C \frac{g(F)}{Q(\eta, \lambda)} \, d\lambda \right] \tag{2.38}$$

where the path C shall enclose all roots of the denominator, pass through the point $F = 0$ or $\lambda_0 = -F_0/t$, and shall not cross the parallel to the positive imaginary axis through that point. (See Fig. 5.)

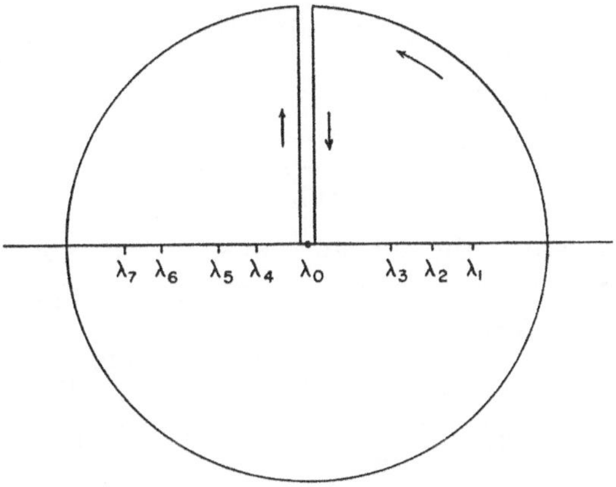

Figure 5

Since all roots of $Q(\eta, \lambda) = 0$ are real, we have for Im $\lambda = \mu$

$$| Q(\eta, \lambda) | = \prod_{k=1}^{m} | \lambda - \lambda_k(\eta) | \geqq | \mu^m | \tag{2.39}$$

Let more precisely α be the highest multiplicity of any root for $| \eta | = 1$. Then there exists a number δ such that a δ-neighborhood of any point λ contains no more than α roots. We have in this case

$$(2.40)\quad |Q(\eta,\lambda)| \geqq \delta^{m-\alpha}\mu^{\alpha};\quad \left|\frac{Q_\lambda(\eta,\lambda)}{Q(\eta,\lambda)}\right| = \left|\sum_{k=1}^{m}\frac{1}{\lambda-\lambda_k}\right| \leqq \frac{m}{|\mu|}$$

for all η on Ω_η and all λ with $|\mathscr{I}m\,\lambda| = |\mu|$.

We assume now that the integer q entering in the definition of g is such that $q \geqq \alpha + 1$. This is certainly the case for $q \geqq m + 1$. We take as path of integration C a circle of radius M about the point $\lambda = \lambda_0 = -F_0/t$ slit along the parallel to the positive imaginary axis through that point. M should be so large that the circle contains all roots of the denominator, e.g.

$$M = |\lambda_0| + 2 \underset{\substack{|\eta|=1\\ k=1,\ldots,m}}{\text{Maximum}} |\lambda_k(\eta)|$$

Along the slit

$$F^{m-1+q} = t^{m-1+q}(\lambda - \lambda_0)^{m-1+q} = t^{m-1+q}(i\mu)^{m-1+q}.$$

It follows from (2.40) that

$$\frac{F^{m-1+q}}{Q(\eta,\lambda)}$$

is continuous along the slit since $m - 1 + q > \alpha$.

We proceed as in the case of the wave equation. We can write R in the form

$$R = \mathscr{R}e\left[\frac{1}{2\pi i}\oint_{|\lambda-\lambda_0|=M}\frac{g(F)}{Q(\eta,\lambda)}\,d\lambda - c\int_{\lambda_0}^{\lambda_0+iM}\frac{F^{m-1+q}}{Q(\eta,\lambda)}\,d\lambda\right]$$

where c denotes the constant

$$c = -1/(m-1+q)!(-2\pi i)^n.$$

In order to form derivatives of R we put R into the form

$$R = \mathscr{R}e\left[ct^{m-1+q}\left(\frac{1}{2\pi i}\oint_{|\lambda|=M}\frac{\lambda^{m-1+q}\left(\frac{\pi i}{2}+\log(\lambda t)\right)}{Q(\eta,\lambda+\lambda_0)}\,d\lambda - \int_0^{iM}\frac{\lambda^{m-1+q}\,d\lambda}{Q(\eta,\lambda+\lambda_0)}\right)\right]$$

The partial derivative of R with respect to λ_0 can be computed by differentiating under the integral sign and then integrating by parts. The resulting expression will not contain any contributions from the limits of integration, and will be identical with what would be obtained directly by differentiating (2.38) with $F = (\lambda - \lambda_0)t$ under the integral sign, disregarding the variability of C.

It follows that one can form the derivatives of R with respect to x or t of orders $\leqq m$ by direct differentiation under the integral sign and that the resulting expressions will be continuous in x, y, t, η. It is to be observed that for $t \to +0$ the path C with its center at the point $\lambda_0 = -F_0/t$ may become unbounded. However for large λ_0 we can always deform the path into a circle of bounded radius about the roots of the denominator, which would not have to pass through λ_0.

It is found then that

$$L[R] = \mathcal{R}e \frac{1}{2\pi i} \oint_C g^{(m)}(F) d\lambda = 0,$$

since $g^{(m)}(F)$ has no singularities inside C. Similarly

$$\frac{\partial^{m-1} R}{\partial t^{m-1}} = \mathcal{R}e \frac{1}{2\pi i} \oint \frac{\lambda^{m-1} g^{(m-1)}(F)}{Q(\eta, \lambda)} d\lambda.$$

Since the path of integration can always be restricted to a bounded portion of the λ-plane, and since $\lim_{t \to 0} F = F_0$ uniformly for bounded λ, we find for $n + q$ even

$$\begin{aligned}\left(\frac{\partial^{m-1} R}{\partial t^{m-1}}\right)_{t=0} &= \mathcal{R}e \frac{g^{(m-1)}(F_0)}{2\pi i} \oint \frac{\lambda^{m-1} d\lambda}{Q(\eta, \lambda)} \\ &= \mathcal{R}e\, g^{(m-1)}((x-y)\cdot\eta) \\ &= \begin{cases} \dfrac{|(x-y)\cdot\eta|^q}{4q!\,(2\pi i)^{n-1}} & \text{for odd } n \\[2ex] \dfrac{-[(x-y)\cdot\eta]^q (\log|(x-y)\cdot\eta| + c')}{q!\,(2\pi i)^n} & \text{for even } n \end{cases}\end{aligned}$$

wit a certain constant c'. (See formula (2.15c).) Similarly the first $m - 2$ derivatives of R with respect to t vanish for $t = 0$.

From this point onwards the solution of the initial value

problem (2.1), (2.2) is obtained exactly as in the strictly hyperbolic case. We define

$$(2.41)\quad Z(x-y,\, t) = \int_{\Omega_\eta} \mathrm{R}\, d\omega_\eta$$

$$= \mathcal{R}e \int_{\Omega_\eta} d\omega_\eta \frac{c}{2\pi i} \oint_C \frac{F^{m-1+q}\left(\frac{\pi i}{2} + \log F\right)}{Q(\eta,\, \lambda)}\, d\lambda,$$

where q must exceed the highest multiplicity α of any root of the characteristic equation, and $q+n$ is even. Then for f in C_{n+q}

$$(2.42)\qquad u(x,\, t) = (\Delta_x)^{(n+q)/2} \int f(y)\, Z(x-y,\, t)\, dy$$

will be a solution of class C_m of the initial value problem.

For $m-n > \alpha$ we can carry out all differentiations in the expression for u under the integral sign, and arrive at formula (2.21), where

$$(2.43)\quad (m-n-1)!\, K(x-y,\, t)$$

$$= \mathcal{R}e \left[\frac{1}{(-2\pi i)^{n+1}} \int_{\Omega_\eta} d\omega_\eta \oint_C \frac{F^{m-n-1} \log F}{Q(\eta,\, \lambda)}\, d\lambda\right].$$

Here the path C used in the contour integral consists of a circle of radius M about the point $\lambda_0 = -F_0/t$ and two line segments along the slit from λ_0 to $\lambda_0 + iM$. For large M the contribution of the circle becomes negligible, and the contour integral can then be replaced by

$$-2\pi i \int_{\lambda_0}^{\lambda_0 + i\infty} \frac{F^{m-n-1}\, d\lambda}{Q(\eta,\, \lambda)}.$$

Putting $\lambda = \lambda_0 + i\mu/t$ this expression can be written

$$-2\pi i^{m-n+1}\, t^{m-1} \int_0^\infty \frac{\mu^{m-n-1}\, d\mu}{Q(t\eta,\, i\mu - F_0)}.$$

Hence

(2.44) $(m-n-1)!\,K(x-y,\,t)$

$$= \mathcal{R}e \frac{i^{m-n} t^{m-1}}{(-2\pi i)^n} \int_{\Omega_\eta} d\omega_\eta \int_0^\infty \frac{\mu^{m-n-1}\, d\mu}{Q(t\eta,\, i\mu - (x-y)\cdot\eta)}.$$

A similar reduction can be given for more general m and n, provided $m \geqq 2 + \alpha$ for odd n or $m \geqq 3 + \alpha$ for even n. In that case one can pull a sufficient number of delta operators in (2.42) under the integral sign to lower the exponent of F in the contour integral below $m-1$. The contour integral can then again be replaced by one taken along the infinite slit with no logarithms in the integrand. [28]

[28] Formulae for the elementary solution which are free of logarithms even in the case of an even number of dimensions have been given by Petrovskii [1]. See Leray [1], pp. 83 et seq.

CHAPTER III

The Fundamental Solution of a Linear Elliptic Differential Equation with Analytic Coefficients[29]

Definition of a fundamental solution

Denote by $L[u]$ a differential operator of the form

$$L[u] = \sum_{k=0}^{m} \sum_{\substack{i_1,\ldots,i_k \\ =1,\ldots,n}} A_{i_1\ldots i_k}(x) \frac{\partial^k u}{\partial x_{i_1} \ldots \partial x_{i_k}}. \tag{3.1}$$

A function $K = K(x, y) = K(x_1, \ldots, x_n, y_1, \ldots, y_n)$ will be called a *fundamental solution* of L, if for every function $f(x)$, which is sufficiently regular and vanishes outside a bounded set,

$$L\left[\int K(x, y) f(y) dy\right] = f(x). \tag{3.2}$$

Symbolically this amounts to $K(x, y)$ being a solution of the inhomogeneous differential equation

$$L[K(x, y)] = \delta(x - y) \tag{3.3}$$

where δ denotes the so-called *Dirac function.* [30]

For the Laplace operator $L = \Delta_x$ we have from Poisson's equation (1.8), (1.9) that a fundamental solution is given by

$$K(x, y) = \begin{cases} \dfrac{|x - y|^{2-n}}{(2 - n)\omega_n} & \text{for } n > 2 \\ \dfrac{1}{2\pi} \log |x - y| & \text{for } n = 2. \end{cases} \tag{3.4}$$

More generally, if L is the iterated Laplacean $(\Delta_k)^r$ a funda-

[29] This chapter is based on John [5], [7]. For elliptic equations in general see Miranda]1[.

[30] See Schwartz [2], vol. 2, pp. 66 et seq.

mental solution is given by

$$(3.5)\qquad K(x-y) = \frac{(-1)^r \Gamma\left(\frac{n}{2} - r\right)}{2^{2r}\pi^{n/2}\Gamma(r)} \mid x - y \mid^{2r-n}$$

for all odd n and for even $n > 2r$, and is given by

$$(3.6)\quad K(x, y) = \frac{(-1)^{(n-2)/2}}{2^{2r-1}\pi^{n/2}(r-1)!\left(r - \frac{n}{2}\right)!} |x-y|^{2r-n}\log|x-y|$$

for even $n \leqq 2r$. (See formulae (1.9a, b).) For $L = \Delta + \lambda^2$, where λ is constant, the formula

$$K = \frac{1}{4}\left(\frac{\lambda}{2\pi \mid x - y \mid}\right)^{\frac{n-2}{2}} N_{\frac{n-2}{2}}(\lambda \mid x - y \mid)$$

represents a fundamental solution (N_ν = "Neumann function" or "Bessel function of the second kind").

We shall restrict ourselves to the case, where L is an *elliptic* operator.[31] This means that the *characteristic form* belonging to L is definite:

$$(3.7)\qquad Q(x, \xi) = \sum_{\substack{i_1, \ldots, i_m \\ = 1, \ldots, n}} A_{i_1 \ldots i_m}(x)\xi_{i_1}\xi_{i_2} \ldots \xi_{i_m} \neq 0 \qquad \text{for real } \xi \neq 0$$

and all x in question. The order m is necessarily even for elliptic L. We assume moreover that the coefficients $A_{i_1 \ldots i_k}(x)$ are *analytic functions* in a neighborhood N of some point. Without restriction of generality we can assume that point to be the origin.

The problem of finding a fundamental solution of L is essen-

[31] Fundamental solutions in the sense defined here can in fact also exist for non-elliptic equations. Thus for a strictly hyperbolic equation of order $m \geqq n + 1$ a fundamental solution is given by

$$K(x, t, y, \tau) = \begin{cases} K(x - y, t - \tau) & \text{for } t > \tau \\ 0 & \text{for } t < \tau \end{cases}$$

where $K(x - y, t)$ is the function given by (2.22), (2.23). For fundamental solutions of more general equations in the form of "distributions" see Ehrenpreis [1], Malgrange [1].

tially equivalent to the problem of finding a solution of the inhomogeneous equation $L[u] = f$ for general f. Since every f can be built up from plane wave functions, it will be sufficient to obtain a solution for an f of the form $f = g(x \cdot \eta)$. By what amounts to "Duhamel's principle" it will actually be sufficient to obtain solutions u for the case where $g = 1$ and u has vanishing Cauchy data on a hyperplane. This reduces the construction of a fundamental solution to the solution of a Cauchy problem for the elliptic equation $L[u] = 1$. The existence of a solution of this Cauchy problem depends strongly on the assumption that the coefficients of L are analytic.

The Cauchy problem

We denote by $v(x, \xi, p)$ the solution of the equation

$$L[v] = 1 \tag{3.8}$$

for which v and all its derivatives of order $\leqq m - 1$ vanish on the hyper-plane $x \cdot \xi = p$. It is known from the theorem of Cauchy and Kowalewski [32] that in a neighborhood of any point x^0 in N lying on the plane there exists a function v satisfying these conditions. (We use here the *elliptic* character of the equation, which assures that every real hyper-plane is non-characteristic, as well as the *analyticity* of the coefficients to apply the theorem.) Moreover v is determined uniquely. Since v depends only on L and the plane, it must be homogeneous of degree 0 in ξ and p.

For our purposes it is important to analyze the dependence of v on ξ and p. For convenience we transform the plane into a fixed one by an orthogonal transformation. This transformation can be chosen *locally* so as to depend continuously on ξ and p. Let η be a unit vector: $|\eta| = 1$. Then the plane $x \cdot \xi = p$ can be transformed into the plane $x' \cdot \eta = 0$ by the orthogonal transformation

$$x' = x + \frac{2x \cdot \xi}{|\xi|}\eta - \frac{x \cdot (\xi + |\xi|\eta)}{|\xi|(|\xi| + \xi \cdot \eta)}(\xi + |\xi|\eta) - \frac{p}{|\xi|}\eta, \tag{3.9}$$

[32] See Courant-Hilbert [1], vol. II, pp. 39 et seq.

as is verified immediately, provided ξ does not have the direction of $-\eta$. (This is just a rotation in a 2-dimensional plane perpendicular to ξ and η.) The inverse transformation is given by the formula

$$(3.10)\quad x = x' + \frac{2x'\cdot\eta}{|\xi|}\xi - \frac{x'\cdot(\xi+|\xi|\eta)}{|\xi|(|\xi|+\xi\cdot\eta)}(\xi+|\xi|\eta) + \frac{p}{|\xi|^2}\xi$$

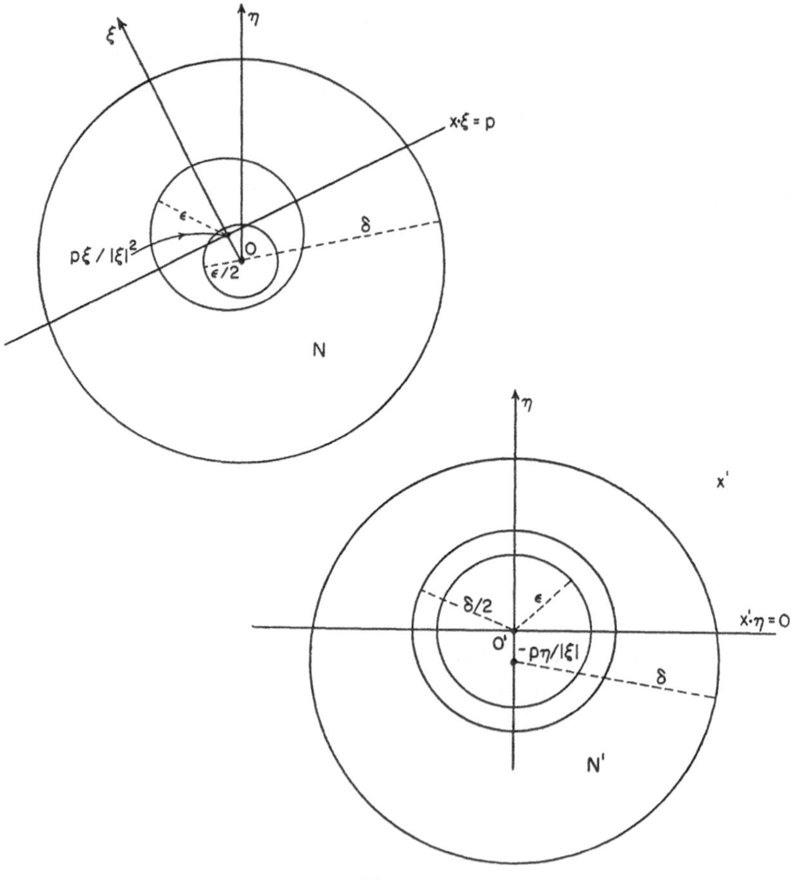

Figure 6

Let now the neighborhood N of the origin in x-space, in which the coefficients of L are analytic, be a sphere of radius δ about the origin. (See Fig. 6.) Moreover restrict ξ to vectors with $|\xi - \eta| < 1/2$ and p to values with $|p| < \delta/4$. Then

$$|\xi| > 1/2, \quad \left|\frac{-p}{|\xi|}\eta\right| < \delta/2.$$

The image of N under the transformation (3.9) is a sphere N' of radius δ about the point $x' = -p\eta/|\xi|$. That latter sphere certainly contains a sphere of radius $\delta/2$ about the origin of x'-space.

Under the substitution (3.10) the coefficients $A(x)$ of L go over into functions $A'(x', \xi, p)$ (for η held fixed). These A' are then analytic functions of x', ξ, p for

$$(3.11) \qquad |x'| < \delta/2, \quad |\xi - \eta| < 1/2, \quad |p| < \delta/4.$$

The operator L goes over into an operator L' acting on functions of x', with coefficients that depend analytically on x' and on the parameters ξ, p in the domain (3.11). The coefficients can be expanded into convergent power series with respect to the x'_i, $\xi_i - \eta_i$, p for $|x'|$, $|\xi - \eta|$, $|p|$ sufficiently small. The Cauchy problem for v goes over into the problem of finding a solution v' of $L'[v'] = 1$, which vanishes with all its derivatives of orders $\leqq m - 1$ on the fixed plane $x' \cdot \eta = 0$. By the theorem of Cauchy and Kowalewski there exists a number $\varepsilon = \varepsilon(\eta) < \delta/4$ and a solution $v' = v'(x', \xi, p)$ of the Cauchy problem given by a convergent power series in the x'_i, $\xi_i - \eta_i$, p for

$$(3.12) \qquad |x'| < \varepsilon, \quad |\xi - \eta| < \varepsilon, \quad |p| < \varepsilon,$$

and hence analytic in x', ξ, p in that region. Transforming back we see that there exists a solution $v = v(x, \xi, p)$, which is analytic in its arguments for

$$\left|x - \frac{p}{|\xi|^2}\xi\right| < \varepsilon, \quad |\xi - \eta| < \varepsilon, \quad |p| < \varepsilon,$$

and hence also for

$$(3.13) \qquad |x| < \varepsilon/2, \quad |p| < \varepsilon/4, \quad |\xi - \eta| < \varepsilon$$

Now η was an arbitrary point of Ω_η. About each of those η we have a sphere $|\xi - \eta| < \varepsilon(\eta)$. A finite number of these solid spheres, say those with centers $\eta^1, \ldots, \eta^s$, will cover some ε'-neighborhood of Ω_η completely. Let $\varepsilon'' = \underset{i=1,\ldots,s}{\text{Minimum}}\, \varepsilon(\eta^i)$. Then $v(x, \xi, p)$ is defined and analytic in its arguments for

$$|x| < \varepsilon''/2, \ |p| < \varepsilon''/4, \ 1 - \varepsilon' < |\xi| < 1 + \varepsilon'.$$

We can assume here that $\varepsilon' < 1$. Then $v(x, \xi, p)$ is analytic for

$$|x| < \varepsilon''/8, \ |p/|\xi|| < \varepsilon''/8, \ 1 - \varepsilon' < |\xi| < 1 + \varepsilon'.$$

The last restriction can be omitted, since v is homogeneous of degree 0 in ξ and p. Writing now ε for $\varepsilon''/8$ we find that the solution of the Cauchy problem is defined and analytic for

$$|x| < \varepsilon, \quad |p/|\xi|| < \varepsilon; \tag{3.14}$$

i.e. for every plane intersecting the sphere of radius ε about the origin $v(x, \xi, p)$ is defined in all points of that sphere, solves the Cauchy problem there, and is analytic in x, ξ, p.

Since $v'(x', \xi, p)$ and all its x'-derivatives of order $\leqq m - 1$ vanish for $x' \cdot \eta = 0$, we can put v' into the form

$$v' = (x' \cdot \eta)^m w'(x', \xi, p),$$

where w' is analytic in its arguments. (This is obvious, when η is one of the coordinate vectors, and follows then for general η). Consequently v can be put into the form

$$v(x, \xi, p) = (x \cdot \xi - p)^m w(x, \xi, p), \tag{3.15}$$

where w is analytic in x, ξ, p in (3.14). Substituting this expression for v into the differential equation $L[v] = 1$, we find that on the hyperplane $x \cdot \xi = p$

$$m!\,Q(x, \xi)\,w(x, \xi, x \cdot \xi) = 1. \tag{3.16}$$

[It may be mentioned [33] that in the special case where the

[33] See John [5], pp. 281—3.

coefficients of the highest order terms in L are constant and where all other coefficients are entire functions (i.e. represented by a power series that converges everywhere) the solution $v(x, \xi, p)$ will also be an entire function of x.]

Solution of the inhomogeneous equation with a plane wave function as right hand side

Let $g(s)$ be a continuous function of s. We introduce the expression

$$V(x, \xi, p) = -\int_0^{x\cdot\xi - p} v_p(x, \xi, t + p) g(t)\, dt, \tag{3.17}$$

which is defined for x, ξ, p satisfying (3.14). Here $v_p(x, \xi, p)$ is a solution of $L[v_p] = 0$, which vanishes with its x-derivatives of orders $\leqq m - 2$ for $p = x \cdot \xi$, and which is such that

$$\begin{aligned} v_{px_{i_1}\cdots x_{i_{m-1}}}(x, \xi, x\cdot\xi) &= -\, m!\, \xi_{i_1} \cdots \xi_{i_{m-1}} w(x, \xi, x\cdot\xi) \\ &= -\, \xi_{i_1} \cdots \xi_{i_{m-1}} / Q(x, \xi) \end{aligned}$$

by (3.15), (3.16). It follows then from (3.17) that all x-derivatives of V of order $\leqq m$ exist and are continuous in x, ξ, p, and that

$$V_{x_{i_1}\cdots x_{i_k}}(x, \xi, p) = -\int_0^{x\cdot\xi - p} v_{px_{i_1}\cdots x_{i_k}}(x, \xi, t+p) g(t)\, dt$$

for $k \leqq m - 1$, and

$$\begin{aligned} V_{x_{i_1}\cdots i_m}(x, \xi, p) &= \xi_{i_1} \cdots \xi_{i_m} g(x\cdot\xi - p)/Q(x, \xi) \\ &\quad - \int_0^{x\cdot\xi - p} v_{px_{i_1}\cdots x_{i_m}}(x, \xi, t+p) g(t)\, dt \end{aligned}$$

Thus $V(x, \xi, p)$ is a solution of

$$L[V] = g(x\cdot\xi - p) \quad \text{(Duhamel's principle)}. \tag{3.18}$$

The fundamental solution

We now choose for $g(s)$ [similarly as in (2.15)] the function

$$g(s) = \mathscr{Re}\left[\frac{-s^k \log(-is)}{k!(2\pi i)^n}\right] \tag{3.19}$$

$$= \begin{cases} \dfrac{|s|^k}{4k!(2\pi i)^{n-1}} & \text{for odd } n \\ -\dfrac{s^k \log|s|}{k!(2\pi i)^n} & \text{for even } n \end{cases}$$

where k is a positive integer and $k + n$ is even. One can take $k = 1$ for odd n and $k = 2$ for even n.

With this function g and the corresponding V given by (3.17) we form

$$W(x, y) = \int_{\Omega_\xi} V(x, \xi, y \cdot \xi) d\omega_\xi \tag{3.20}$$

$$= -\int_{\Omega_\xi} d\omega_\xi \int_0^{(x-y)\cdot\xi} v_p(x, \xi, t + y \cdot \xi)\, g(t) dt$$

Then $W(x, y)$ is defined for x and y in an ε-neighborhood of the origin, is of class C_m in x, and satisfies the equation

$$L[W] = \int_{\Omega_\xi} g((x - y) \cdot \xi)\, d\omega_\xi. \tag{3.21}$$

We shall prove that $W(x, y)$ is analytic in x and y for

$$|x| \leqq \frac{\varepsilon}{6}, \quad |y| \leqq \frac{\varepsilon}{6}, \quad x \neq y. \tag{3.22}$$

Let $x - y = r\zeta$, where $|\zeta| = 1$, $r \geqq 0$. Integrating the inner integral in (3.20) by parts, introducing a new variable of integration s by $t = rs$, and using (3.15) we obtain

$$W(x, y) = \int_{\Omega_\xi} d\omega_\xi \int_0^{r\zeta\cdot\xi} v(x, \xi, t + y\cdot\xi) g'(t)dt$$

$$= r^{m+1} \int_{\Omega_\xi} d\omega_\xi \int_0^{\zeta\cdot\xi} (\zeta\cdot\xi - s)^m w(x, \xi, rs + y\cdot\xi)\, g'(rs)ds.$$

We can make the limits of integration locally independent of x and y by a suitable orthogonal substitution. Let η be a unit vector and let the vector ζ be restricted to the half-space $\zeta\cdot\eta > 0$. In analogy with (3.10) we introduce a new variable of integration ξ' instead of ξ by the formula

$$(3.23)\quad \xi = \xi' + \frac{2\xi'\cdot\eta}{|\zeta|}\zeta - \frac{\xi'\cdot(\zeta + |\zeta|\eta)}{|\zeta|\,(|\zeta| + \zeta\cdot\eta)}(\zeta + |\zeta|\eta) = T(\xi', \zeta).$$

For any ζ this is an orthogonal substitution: $|T(\xi', \zeta)| = |\xi'|$. In addition $\zeta\cdot\xi = |\zeta|\xi'\cdot\eta$. The expression for W becomes for $|\zeta| = 1$

$$(3.24)\qquad W(x, y) =$$

$$r^{m+1} \int_{\Omega_{\xi'}} d\omega_{\xi'} \int_0^{\xi'\cdot\eta} (\xi'\eta - s)^m w(x, T(\xi', \zeta), rs + y\cdot T(\xi', \zeta)) g'(rs)ds$$

where the limits do not depend on x and y, at least for $x - y$ restricted to the half-space $(x - y)\cdot\eta > 0$. The function $w(x, \xi, p)$ had been shown to be an analytic function of its arguments, when those arguments are real and satisfy $|x| < \varepsilon$, $|p|/|\xi| < \varepsilon$. The function $T(\xi', \zeta)$ is analytic in ζ by (3.23) for $\zeta\cdot\eta > 0$, ζ real, and satisfies $|T| = 1$ for ξ' on $\Omega_{\xi'}$. Moreover for the s in the interval of integration

$$|rs + y\cdot T(\xi', \zeta)| \leqq |r\xi'\cdot\eta| + |y\cdot T(\xi', \zeta)| \leqq r + |y| \leqq \varepsilon$$

for $|r| < \varepsilon/2$, $|y| < \varepsilon/2$. It follows that

$$(3.25)\qquad w(x, T(\xi', \zeta), rs + y\cdot T(\xi', \zeta))$$

is a regular analytic function [34] of x, y, r, ζ, s, ξ' for real values of those variables in the set

$$(3.26) \quad |x| \leqq \varepsilon/3, \ |y| \leqq \varepsilon/3, \ |r| \leqq \varepsilon/3, \ |\zeta| = 1, \ |s| \leqq 1, \quad |\xi'| = 1, \ \zeta \cdot \eta \geqq \tfrac{1}{2}.$$

The closed set (3.26) can be covered completely by a finite number of neighborhoods of points of the set, in each of which w is represented by a convergent power series. It follows that we can choose a *universal* ε'' such that for every point x^0, y^0, r^0, ζ^0, s^0, ξ'^0 of (3.26) the function w is represented by a series in the $x_i^0 - x_i$, $y_i^0 - y_i$, $\zeta_i^0 - \zeta_i$, $\xi_i'^0 - \xi_i'$, $r^0 - r$, $s^0 - s$, which converges absolutely for

$$|x - x^0| \leqq \varepsilon'', \ |y - y^0| \leqq \varepsilon'', \ |r - r^0| \leqq \varepsilon'', \ ^0|s - s^0| \leqq \varepsilon'', \quad |\zeta - \zeta^0| \leqq \varepsilon'', \ |\xi' - \xi'^0| \leqq \varepsilon''.$$

We can write the series as a series in the $x_i - x_i^0$, $y_i - y_i^0$, $r - r^0$, $\zeta_i - \zeta_i^0$ alone, with coefficients, which depend on ξ' and s. The coefficients will then be given by power series in the $\xi_i' - \xi_i'^0$, $s - s^0$, and hence will be analytic functions of ξ' and s, and in particular will be continuous for $|s - s^0| \leqq \varepsilon''$, $|\xi' - \xi'^0| \leqq \varepsilon''$. Since ξ'^0 and s^0 are arbitrary points of the set $|\xi'| = 1$, $|s| \leqq 1$, we see that there exists an ε'', such that for x^0, y^0, r^0, ζ^0 in the set

$$(3.27) \quad |x| \leqq \varepsilon/3, \ |y| \leqq \varepsilon/3, \ |r| \leqq \varepsilon/3, \ |\zeta| = 1, \ \zeta \cdot \eta \geqq \tfrac{1}{2}$$

the function w is represented by an absolutely convergent power series in the $x_i - x_i^0 \cdot y_i - y_i^0$, $r - r^0$, $\zeta_i - \zeta_i^0$ for

$$(3.28) \quad |x - x^0| \leqq \varepsilon'', \ |y - y^0| \leqq \varepsilon'', \ |r - r^0| \leqq \varepsilon'', \ |\zeta - \zeta^0| \leqq \varepsilon''.$$

[34] Analyticity of a function $F(y_1, \ldots, y_k)$ at a real point $(y_1^0, \ldots, y_k^0) = P^0$ is always used here in the sense that the function is representable by an absolutely convergent power series in the $y_i - y_i^0$ in a real ε'-neighborhood of P^0, i.e. for

$$|y_i - y_i^0| \leqq \varepsilon' \quad \text{for } i = .\ 1, \ldots, k.$$

(This implies that the function can also be defined as an analytic function in a *complex* neighborhood of P^0, though this fact will not be used here.) The radius ε' will in general depend on P^0. However if P^1 is a point of the ε'-neighborhood of P^0, the power series of F about the point P^1 will converge in the largest neighborhood with center P^1, which is contained in the ε'-neighborhood of P^0.

The coefficients of this series are continuous functions of a s and ξ' for $|\xi'| = 1$, $|s| \leqq 1$.

For positive r the function $g'(rs)$ is given by

$$(3.29)\quad g'(rs) = \begin{cases} \dfrac{r^{k-1}s^{k-1}\operatorname{sign} s}{4(k-1)!\,(2\pi i)^{n-1}}, & \text{if } n \text{ is odd} \\ \dfrac{-r^{k-1}s^{k-1}}{(k-1)!\,(2\pi i)^{n}}\left(\log r + \log|s| + \dfrac{1}{k}\right), & \text{if } n \text{ is even.} \end{cases}$$

We substitute this expression for g' into the integral (3.24) and substitute the power series for w. Integrating term by term there results for positive r an expression for W of the form

$$(3.30)\quad W(x, y) = r^{m+k}(A(x, y, r, \zeta) + B(x, y, r, \zeta)\log r),$$

where $B = 0$ for odd n. The functions A and B are given by absolutely convergent power series in the set (3.28). It follows that they are analytic at the arbitrary point x^0, y^0, r^0, ζ^0 of the set (3.27). Hence $W(x, y)$ is an analytic function of x, y, r, ζ in the set (3.27) for $r > 0$. Taking $r = |x - y|$, $\zeta = (x - y)/r$, we see that $W(x, y)$ is analytic in x and y for

$$(3.31)\quad |x| \leqq \frac{\varepsilon}{6},\quad |y| \leqq \frac{\varepsilon}{6},\quad x \neq y,\quad \frac{(x-y)\cdot\eta}{|x-y|} \geqq \frac{1}{2}.$$

Since η was an arbitrary unit vector, it follows that $W(x, y)$ is analytic in x and y in the set (3.22).

If we substitute for r, ζ in (3.30) their values $r = |x - y|$, $\zeta = (x - y)/|x - y|$, and differentiate with respect to x_i, we find for W_{x_i} an expression of the same form

$$W_{x_i} = r^{m+k-1}(A'(x, y, r, \zeta) + B'(x, y, r, \zeta)\log r)$$

where

$$A'(x, y, r, \zeta) = (m+k)A\zeta_i + rA_{x_i} + B\zeta_i + A_{\zeta_i} - \sum_k A_{\zeta_k}\zeta_i\zeta_k + rA_r\zeta_i$$

$$B'(x, y, r, \zeta) = (m+k)B\zeta_i + rB_{x_i} + B_{\zeta_i} - \sum_k B_{\zeta_k}\zeta_i\zeta_k + rB_r\zeta_i.$$

A' and B' have the same regularity properties as A and B. An expression of the same type can be given for W_{y_i}.

Let W' denote any derivative of W with respect to the x_i and y_i of (total) order ν. It can be represented in the form (3.30) with k replaced by $k - \nu$. Thus W' will be analytic in the set (3.22) and

$$|W'|r^{-m-k+\nu} \text{ for odd } n$$

$$|W'|r^{-m-k+\nu}/\log r \text{ for even } n$$

will stay bounded near $x = y$.

We form in particular the expression

$$K(x, y) = (\Delta_y)^{(n+k)/2} W(x, y), \tag{3.32}$$

which will turn out to be the desired fundamental solution of L. Then K is analytic in x and y in the set (3.22). The singularity of K at $x = y$ is such that $r^{n-m}K$ stays bounded for odd n and becomes at most logarithmically infinite for even n. More generally for any derivative K' of order ν of K the expression $r^{n-m+\nu}K'$ is bounded, respectively becomes logarithmically infinite.

By (3.19), (3.21), (1.6), (1.7) $L[W]$ is of the form

$$r^k(c_1 + c_2 \log r)$$

with certain constants c_1, c_2, where $c_2 = 0$ for odd n. It follows that in the set (3.22)

$$L[K] = (\Delta_y)^{(n+k)/2} L[W] = 0. \tag{3.33}$$

In order to verify the property (3.2) of K we take a function $f(x)$ of class C_1, which vanishes outside an $\varepsilon/6$-neighborhood of the origin. Let

$$u(x) = \int K(x, y)f(y)dy = \int f(y)(\Delta_y)^{(n+k)/2} W(x, y)dy.$$

Then u can be written in the form

$$u(x) = -\sum_i \int f_{y_i}(y) \frac{\partial}{\partial y_i} (\Delta_y)^{(n+k-2)/2} W(x, y)dy.$$

Since the derivatives of $W(x, y)$ of order $\leqq m + k + n - 1$

with respect to x or y are still absolutely integrable, we have then

$$L[u] = -\sum_i \int f_{y_i}(y) \frac{\partial}{\partial y_i} (\Delta_y)^{(n+k-2)/2} L[W(x, y)]\, dy.$$

Using the fact that by (3.21) the function $L[W(x, y)]$ only depends on the difference $x - y$, it follows that

$$\begin{aligned} L[u] &= -\sum_i \int f_{y_i}(y) \frac{\partial}{\partial y_i} (\Delta_x)^{(n+k-2)/2} L[W(x, y)]\, dy \\ &= -(\Delta_x)^{(n+k-2)/2} \sum_i \int f_{y_i}(y) \frac{\partial}{\partial y_i} L[W(x, y)]\, dy \\ &= (\Delta_x)^{(n+k-2)/2} \int f(y)\, \Delta_y L[W(x, y)]\, dy \\ &= (\Delta_x)^{(n+k)/2} \int f(y)\, L[W(x, y)]\, dy \\ &= (\Delta_x)^{(n+k)/2} \int f(y) \left(\int_{\Omega_\xi} g((x-y)\cdot \xi)\, d\omega_\xi \right) dy. \end{aligned}$$

Relations (1.10), (1.11) then have as a consequence that

$$L[u] = f(x). \tag{3.34}$$

Hence the function $K(x, y)$ constructed is a fundamental solution.

We can generalize (3.34) slightly to the case, where f does not have to vanish outside a bounded set. Let D be a domain contained in the $\varepsilon/6$-neighborhood of the origin, in which K is defined. Let $f(y)$ be a function, which is of class C_1 in D and is continuous in the closure of D. Then (3.34) is satisfied for

$$u = \int_D f(y) K(x, y) dy \tag{3.35}$$

at every point x of D.

For the proof we take a point x of the open set D. We can decompose $f(y)$ into a sum $f_1(y) + f_2(y)$, where $f_1(y)$ is continuous in the closure of D and vanishes in a neighborhood of the point x, and where $f_2(y)$ is of class C_1 for all y and vanishes outside D.

If $u_1(x)$ and $u_2(x)$ denote the integrals corresponding to f_1 and f_2, we have

$$L[u_2(x)] = f(x)$$

by what has been proved, and

$$L[u_1(x)] = 0$$

as a consequence of (3.33).

The function $K(x, y)$ has another property, which also could be considered as an interpretation of the symbolic equation (3.3). Let R be a region with boundary B, which is sufficiently regular to permit application of the divergence theorem. If we take the expression

$$\int_R u(x)L[v(x)]dx$$

for arbitrary functions u and v of class C_m, and integrate it m-times by parts, so as to remove all derivatives of v from the integrand, we end up with an identity of the type

$$\int_R (u(x)L[v(x)] - v(x)\bar{L}[u(x)])\,dx = \int_B M[u, v]\,dS_x \tag{3.36}$$

Here $\bar{L}$ is the so-called "adjoint" differential operator to L, which can be put into the form

$$\bar{L}[u] = \sum_{k=0}^{m} (-1)^k \sum_{\substack{i_1, \ldots, i_k \\ =1, \ldots, n}} \frac{\partial^k}{\partial x_{i_1} \ldots \partial x_{i_k}} [A_{i_1 \ldots i_k} u]. \tag{3.37}$$

The expression $M[u, v]$ is a bi-linear differential operator. It is linear and homogeneous in u and its derivatives of orders $\leqq m - 1$, and also in v and its derivatives of order $\leqq m - 1$, and in the direction cosines of the normal of B. The total order of derivatives of u and v occurring in each term of $M[u, v]$ is at most $m - 1$. Formula (3.36) is known as *Green's identity* tor the operator L.

We apply Green's identity to a function u of class C_m and to

$v(x) = W(x, y)$, which also will be of class C_m, at least for sufficiently large k. Then

$$\int_R (u(x)L[W(x,y)] - W(x,y)\bar{L}[u(x)])\,dx = \int_B M[u(x), W(x,y)]dS_x.$$

Here it is assumed that R is contained in the $\varepsilon/6$-neighborhood of the origin, and that y is an interior point of R. We apply the operator $(\Delta_y)^{(n+k)/2}$ to both sides of this equation. Remembering that by (3.21), (1.10), (1.11) [35]

$$(\Delta_y)^{(n+k)/2} \int_R u(x)\,L[W(x, y)]dx = u(y),$$

and that $W(x, y)$ is regular for x on B and y in the interior of R, we find the desired identity

$$(3.38) \quad u(y) = \int_R K(x,y)\bar{L}[u(x)]dx + \int_B M[u(x), K(x,y)]dS_x.$$

Let in particular $u(x)$ be a solution of the elliptic equation $\bar{L}[u] = 0$. Then the volume integral disappears from (3.38). The integral over B is an analytic function of y for y interior to R, since $K(x, y)$ is analytic in x and y for $x \neq y$. It follows that $u(y)$ is analytic in a neighborhood of the origin. Since $\bar{L}$ is an arbitrary elliptic operator with analytic coefficients, in a neighbourhood of the origin we have the result that *any solution u of class C_m of a linear elliptic equation of order m is analytic at every point of its domain of existence, at which the coefficients are analytic.*

Characterization of the fundamental solution [35a] *by its order of magnitude*

THEOREM: *Let $u(x)$ be a solution of the equation $L[u] = 0$ in a deleted neighborhood of a point z. Assume that for every derivative*

[35] The integrals in (1.10), (1.11) are taken over the whole space instead of over the set R. However we can extend the result to integrals over R for y interior to R in the same way as in the proof of (3.34), (3.35).

[35a] See Miranda [1] pp. 52-53 for the case $m=2$; Bers [1] for equations fo higher order.

$u^{(m-1)}(x)$ *of* u *of order* $m-1$

$$\lim_{x\to z} |x-z|^n u^{(m-1)}(x) = 0 \tag{3.39}$$

Then

$$u(x) = cK(x, z) + w(x),$$

where $K(x, z)$ *is the fundamental solution constructed previously,* c *is a constant, and* $w(x)$ *is a solution of* $L[w] = 0$, *which is regular analytic at the point* $x = z$.

Proof: We first show that relation (3.39) implies that for every derivative $u^{(k)}$ of u of order $k < m-1$

$$\lim_{x\to z} |x-z|^{n-1} u^{(k)}(x) = 0. \tag{3.41}$$

It is sufficient to show this for $k = m-2$, since (3.41) implies a fortiori that

$$\lim_{x\to z} |x-z|^n u^{(k)}(x) = 0$$

and the argument can be repeated. Now by assumption there is a δ-neighborhood of the point z, such that

$$|u^{(m-1)}(x)| \leqq \varepsilon |x-z|^{-n}/\sqrt{n}$$

in that neighborhood. Let x be a point of that neighborhood and let y be the point on the same ray from z, such that $|y-z| = \delta$. Then

$$|u^{(m-2)}(y) - u^{(m-2)}(x)| \leqq \varepsilon \int_{|x-z|}^{|y-z|} s^{-n}\, ds = \varepsilon \frac{|x-z|^{1-n} - \delta^{1-n}}{n-1}$$

Now $u^{(m-2)}(y)$ is bounded on the sphere of radius δ about z. Multiplying the last relation with $|x-z|^{n-1}$ and letting x tend to z, we obtain

$$\operatorname*{lim\,superior}_{x\to z} |x-z|^{n-1} |u^{(m-2)}(x)| \leqq \varepsilon/(n-1).$$

Since ε is arbitrary, relation (3.41) follows for $k = m-2$.

Let now $u(x)$ be a solution of $\bar{L}[u] = 0$ in a deleted neighborhood of the point z. (We interchange here the roles of L and $\bar{L}$ to keep the notation simpler). The Green's identity (3.36)

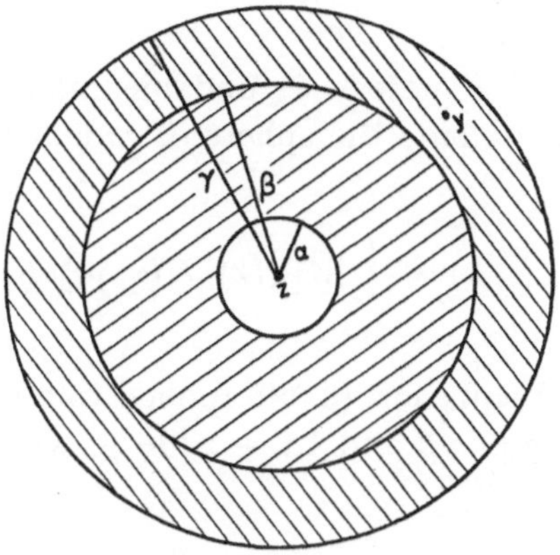

Figure 7

applied to a shell (see Fig. 7) of radii α and β about z yields for any function $v(x)$ of class C_m

$$\int_{\alpha<|x-z|<\beta} u(x)L[v(x)]dx = \int_{|x-z|=\beta} M[u, v]dS_x - \int_{|x-z|=\alpha} M[u, v]dS_x.$$

On the sphere $|x - z| = \alpha$

$$\begin{aligned} v(x) - v(z) &= 0(\alpha) \\ u^{(m-1)}(x) &= o(\alpha^{-n}) \\ u^{(k)}(x) &= o(\alpha^{1-n}) \quad \text{for } 0 \leqq k \leqq m-2. \end{aligned}$$

Hence on that sphere

$$\begin{aligned} M[u(x), v(x)] &= M[u(x), v(z)] + M[u(x), v(x) - v(z)] \\ &= v(z)M[u(x), 1] + o(\alpha^{1-n}). \end{aligned}$$

For $\alpha \to +o$ we then obtain

$$(3.42) \qquad \int_{|x-z|<\beta} u(x)\, L[v(x)]\, dx = \int_{|x-z|=\beta} M[u, v]\, dS_x - cv(z),$$

where

$$c = \lim_{\alpha \to +0} \int_{|x-z|=\alpha} M[u(x), 1]dS_x = \int_{|x-z|=\beta} M[u(x), 1]dS_x - \int_{|x-z|<\beta} u(x)L[1]dx.$$

We choose for $v(x)$ now the fundamental solution $K(x, y)$, where y is a point outside the sphere of radius β about z. Then (3.42) yields

$$cK(z, y) = \int_{|x-z|=\beta} M[u(x), K(x, y)]dS_x.$$

On the other hand equation (3.38) applied to a shell bounded by the spheres $|x - z| = \beta$ and $|x - z| = \gamma$, where $\gamma > |y - z|$, yields

$$u(y) = \int_{|x-z|=\gamma} M[u(x), K(x, y)]dS_x - \int_{|x-z|=\beta} M[u(x), K(x, y)]dS_x.$$

Hence

$$u(y) = \int_{|x-z|=\gamma} M[u(x), K(x, y)]dS_x - cK(z, y) = w(y) - cK(z, y)$$

where $w(y)$ is analytic at the point z, since $K(x, y)$ is analytic in y for $x \neq y$. (We assume that all spheres considered are so small that $K(x, y)$ is defined.)

We can take in particular for $u(x)$ the function $u_0(x) = \overline{K}(x, z)$ where $\overline{K}$ is the fundamental solution constructed for the operator $\overline{L}$. We obtain

$$u_0(y) = w_0(y) - c_0K(z, y).$$

Here the constant c_0 cannot vanish, for otherwise $\overline{K}(x, z)$ considered as a function of x would be regular analytic at $x = z$. This however is impossible, since by (3.38) for arbitrary $v(x)$

$$v(z) = \int_R \overline{K}(x, z)\, L[v(x)]dx + \int_B \overline{M}[v(x), \overline{K}(x, z)]dS_x$$

whereas for $\overline{K}(x, z)$ regular at $x = z$ we would get the same identity from (3.36) with the left hand side replaced by 0.

It follows that

$$u(y) = \frac{c}{c_0}\overline{K}(y, z) + w(y) - \frac{c}{c_0} w_0(y),$$

where w and w_0 are regular at the point z. This completes the proof of the theorem. [36]

The theorem proved states essentially that the fundamental solution $K(x, z)$ has an isolated singularity at $x = z$ which is of the lowest possible order. [37]

Structure of the fundamental solution

In the preceding paragraphs a fundamental solution $K(x, y)$ for a linear elliptic operator L has been constructed, at least for x and y restricted to a neighborhood of a point, where the coefficients of L are analytic. [38] It was found that for x, y restricted to a neighborhood of the origin, or more generally to the neighborhood of a point z at which the coefficients of L are analytic, K can be put into the form

$$K(x, y) = r^{m-n}(A'(x, y, r, \zeta) + B'(x, y, r, \zeta) \log r), \tag{3.43}$$

where $r = |x - y|$, $\zeta = (x - y)/r$. This representation is only local, and hence also not unique. The functions A' and B' still depend on a unit vector η as parameter, and ζ is restricted to $\zeta \cdot \eta \geqq 1/2$. The functions A' and B' are analytic for real values

[36] Condition (3.39) can be replaced by weaker ones without affecting the result (3.40). It can be shown (see p. 156) that (3.39) for a solution u of $L[u] = 0$ is a consequence of the simpler relation

$$\lim_{x \to z} |x - z|^{n-m+1} u(x) = 0.$$

[37] For isolated singularities of higher order see John [5], pp. 293—8.

[38] Using the observation (see p. 49) that the Cauchy problem for L with constant highest coefficients and entire lower coefficients can be solved in the large, one can conclude that for such L the $K(x, y)$ constructed exists and is analytic for all real x, y with $x \neq y$. For the question of existence of fundamental solutions in the large for more general L see John [7]; Browder [2], p. 745; E. E. Levi [1].

of their arguments x, y, r, ζ provided $|x-z|$, $|y-z|$, $|r|$ are sufficiently small and $|\zeta|=1$, $\zeta \cdot \eta \geqq 1/2$. We also have the representation

$$K(x, y) = (\Delta_y)^{(n+k)/2} W(x, y)$$
$$= -(\Delta_y)^{(n+k)/2} \int_{\Omega_\xi} d\omega_\xi \int_0^{(x-y)\cdot\xi} v_p(x, \xi, t + y \cdot \xi) g(t) dt$$

where $g(s)$ is given by (3.19).

For *odd* n we may take $k = 1$. Observing that

$$\Delta_y \int_0^{(x-y)\cdot\xi} v_p(x, \xi, t+y\cdot\xi)\,|t|\,dt = \Delta_y\Big(\text{sign}[(x-y)\cdot\xi] \int_{y\cdot\xi}^{x\cdot\xi} v_p(x, \xi, t)(t-y\cdot\xi)dt\Big)$$
$$= v_p(x, \xi, y \cdot \xi)\ \text{sign}\,[(x-y)\cdot\xi],$$

we find

$$K(x, y) = \frac{-(\Delta_y)^{(n-1)/2}}{4(2\pi i)^{n-1}} \int_{\Omega_\xi} v_p(x, \xi, y \cdot \xi)\ \text{sign}\,[(x - y \cdot \xi] d\omega_\xi. \tag{3.44}$$

The function

$$A'(y + r\zeta, y, r, \zeta) = r^{-m+n} K(y + r\zeta, y) \tag{3.45}$$

is analytic in y, r, ζ for $|y-z|$, $|r|$ sufficiently small and $|\zeta|=1$, $\zeta . \eta \geqq 1/2$. Hence it can be expanded into a series of powers of r

$$\sum_{\alpha=0}^{\infty} c_\alpha(y, \zeta) r^\alpha$$

where the coefficients are analytic in y, ζ. Since the c_α could also be formed by differentiation with respect to r of the right hand side of (3.45), which is independent of η, it follows that the $c_\alpha(y, \zeta)$ are independent of η and are analytic in y, ζ for $|y-z|$ sufficiently small and all ζ with $|\zeta|=1$. Thus for odd n we have an expansion for K of the form

$$K(x, y) = r^{m-n} \sum_{\alpha=0}^{\infty} c_\alpha(y, \zeta) r^\alpha. \tag{3.46}$$

It is easy to give an explicit representation for the coefficient c_0 of the lowest order therm in r. We have from (3.15)

$$v_p(x, \xi, y \cdot \xi) \operatorname{sign} [(x-y) \cdot \xi] =$$
$$= - |(x-y) \cdot \xi|^{m-1} (mw(x, \xi, y \cdot \xi) - (x-y) \cdot \xi w_p(x, \xi, y \cdot \xi)).$$

It is clear that the only contributions of order r^{m-n} arise by neglecting the term with w_p and by putting $x = y$ in $w(x, \xi, y \cdot \xi)$. Using (3.15) we find that the lowest order term of K is given by

$$(3.47) \quad r^{m-n} c_0(y, \zeta) = (\Delta_y)^{(n-1)/2} \int_{\Omega_\xi} \frac{|(x-y) \cdot \xi|^{m-1}}{4(2\pi i)^{n-1}(m-1)!\, Q(y, \xi)} d\omega_\xi.$$

(This is the only term of K in the case, where L only contains derivatives of order m and has constant coefficients. See below, p. 66.)

We turn now to the case of an *even* number n of dimensions. We have from (3.32), (3.20), (3.19) with $k = 2$

$$K(x, y) = (\Delta_y)^{(n+2)/2} \int_{\Omega_\xi} d\omega_\xi \int_0^{(x-y) \cdot \xi} \frac{v_p(x, \xi, t + y \cdot \xi) t^2 \log |t|}{2(2\pi i)^n} dt$$
$$= (\Delta_y)^{n/2} \int_{\Omega_\xi} d\omega_\xi \int_0^{(x-y) \cdot \xi} \frac{v_p(x, \xi, t + y \cdot \xi)(\frac{3}{2} + \log |t|)}{(2\pi i)^n} dt.$$

One verifies easily that the term with 3/2 only contributes a regular solution of the differential equation $L[u] = 0$, and hence can be omitted. Integrating the remainder by parts with respect to t we find

$$K(x, y) = K_1(x, y) + K_2(x, y),$$

where

$$(3.47a) \quad (2\pi i)^n K_1(x, y) = -(\Delta_y)^{n/2} \int_{\Omega_\xi} v(x, \xi, y \cdot \xi) \log |(x-y) \cdot \xi| \, d\omega_\xi$$

$$(3.47b) \quad (2\pi i)^n K_2(x, y) = -(\Delta_y)^{n/2} \int_{\Omega_\xi} d\omega_\xi \int_0^{(x-y) \cdot \xi} \frac{v(x, \xi, t + y \cdot \xi) - v(x, \xi, y \cdot \xi)}{t} dt$$

Here $K_2(x, y)$ is regular analytic in x and y even for $x = y$, so that K_1 contains the singular part of K. (The function K_2 is not necessarily a solution of the differential equation $L[u] = 0$.) Putting $x - y = r\zeta$, we see that the coefficient of $\log r$ in the expression for K_1 is given by

$$(3.48) \qquad w(x, y) = -(2\pi i)^{-n} (\Delta_y)^{n/2} \int_{\Omega_\xi} v(x, \xi, y \cdot \xi) d\omega_\xi$$

$$= -(2\pi i)^{-n} \int_{\Omega_\xi} v^{(n)}(x, \xi, y \cdot \xi) d\omega_\xi$$

where $v^{(n)}(x, \xi, p)$ denotes $\partial^n v(x, \xi, p)/\partial p^n$. It is clear from this representation that $w(x, y)$ is a regular solution of $L[w] = 0$, since the integrand is regular analytic and satisfies that differential equation for all ξ, p. The remaining portion of K_1 and also K_2 can be expanded into a series of powers of r with coefficients which are regular analytic in ζ and y, as in the case of odd n. One finds in this way that for even n the function $K(x, y)$ is of the form

$$(3.49) \qquad K(x, y) = r^{m-n} \sum_{\alpha=0}^{\infty} c_\alpha(y, \zeta) r^\alpha + w(x, y) \log r,$$

where $w(x, y)$ is a regular analytic solution of $L[w] = 0$.

The example of the potential equation in more than 2 dimensions shows that the logarithmic portion w of K can be absent, even in an even number of dimensions. Non-trivial examples of second order differential equations for which the fundamental solution contains no logarithmic portion have been given by Stellmacher [1] in connection with the problem of determining hyperbolic equations, for which Huygens' principle is valid. Let n be an integer > 4. Then we have the example of the equation

$$L[u] = \Delta_x u - \frac{2}{(x_n + B)^2} u = 0$$

(B = arbitrary constant) which has the fundamental solution

$$K(x, y) = \frac{1}{(2-n)\omega_n}\left[|x-y|^{2-n} - \frac{|x-y|^{4-n}}{(n-4)(y_n+B)(x_n+B)}\right].$$

We shall show, however, that $w \equiv 0$ *is only possible, when the order* m *of* L *is less than the number* n *of dimensions (for* n *even).* We have indeed from (3.15) for $n \leqq m$

$$v^{(n)}(x, \xi, y\cdot\xi) = \frac{m!}{(m-n)!}\, r^{m-n}(\zeta\cdot\xi)^{m-n} w(x, \xi, y\cdot\xi) + 0(r^{m-n+1})$$

$$= \frac{m!}{(m-n)!}\, r^{m-n}(\zeta\cdot\xi)^{m-n} w(y, \xi, y\cdot\xi) + 0(r^{m-n+1}),$$

and hence from (3.16), (3.48)

$$(3.50) \qquad -(2\pi i)^n w(x, y) = \frac{r^{m-n}}{(m-n)!}\int_{\Omega_\xi} \frac{(\zeta\cdot\xi)^{m-n}}{Q(y, \xi)}\, d\omega_\xi + 0(r^{m-n+1}).$$

Since $m - n$ is an even number, the integral certainly does not vanish, and hence

$$\lim_{x\to y} r^{n-m} w(x, y) \neq 0.$$

Formulae (3.5), (3.6) for the fundamental solution of the iterated Laplacean illustrate the distinction of the cases $m \geqq n$ and $m < n$. Formula (3.49) shows that for $m < n$ the logarithmic portion, even if present, is outweighed by the other terms, since $w(x, y)$ is bounded.

The fundamental solution for elliptic operators with constant coefficients

Let L have constant coefficients. We can write L in the form

$$(3.51) \qquad L = P\left(\frac{\partial}{\partial x_1}, \ldots, \frac{\partial}{\partial x_n}\right)$$

where P is a polynomial of degree m with constant coefficients. The solution $v(x, \xi, p)$ of $L[v] = 1$ with zero Cauchy data on $x\cdot\xi = p$ is given in this case by

$$v(x, \xi, p) = \frac{1}{2\pi i}\oint_C \frac{e^{(x\cdot\xi-p)\lambda}}{\lambda P(\lambda\xi)}\,d\lambda, \tag{3.52}$$

where C is a path in the complex λ-plane, which encloses all roots of the denominator. Since the coefficient of λ^{m+1} in the polynomial $\lambda P(\lambda\xi)$ is the characteristic form $Q(\xi)$, which does not vanish on Ω_ξ, the roots of the denominator are bounded uniformly in ξ for $|\xi| = 1$. Hence we can choose for C a circle $|\lambda| = M$, where M does not depend on ξ.

We first take up the case of an odd n. In this case formula (3.44) yields

$$K(x, y) = \frac{1}{4(2\pi i)^n}(\Delta_y)^{(n-1)/2}\int_{\Omega_\xi}\operatorname{sign}[(x-y)\cdot\xi]\,d\omega_\xi\oint_{|\lambda|=M}\frac{e^{\lambda(x-y)\cdot\xi}}{P(\lambda\xi)}\,d\lambda. \tag{3.53}$$

In the particular case, where P is homogeneous of degree m, and hence $P = Q$, we can evaluate the contour integral, and obtain

$$K(x, y) = \frac{1}{4(2\pi i)^{n-1}(m-1)!}(\Delta_y)^{(n-1)/2}\int_{\Omega_\xi}\frac{|(x-y)\cdot\xi|^{m-1}}{Q(\xi)}\,d\omega_\xi \tag{3.54}$$

in agreement with (3.47). For $n \leqq m$ this reduces to

$$K(x, y) = \frac{1}{4(m-n)!(2\pi i)^{n-1}}\int_{\Omega_\xi}\frac{[(x-y)\cdot\xi]^{m-n}\operatorname{sign}(x-y)\cdot\xi}{Q(\xi)}\,d\omega_\xi. \tag{3.54a}$$

The contour integral in (3.53) vanishes of order $m - 1$ for $(x - y)\cdot\xi = 0$. It follows that for $n < m$ the differentiations can be carried out under the integral sign, without any contributions from the discontinuity of sign $(x - y)\cdot\xi$. Thus for odd $n < m$

$$K(x, y) = \frac{1}{4(2\pi i)^n}\int_{\Omega_\xi}\operatorname{sign}[(x-y)\cdot\xi]\,d\omega_\xi\oint_{|\lambda|=M}\frac{e^{\lambda(x-y)\cdot\xi}}{P(\lambda\xi)}\lambda^{n-1}\,d\lambda.$$

The integral stays unchanged in form if ξ and λ are replaced by $-\xi$ and $-\lambda$. It follows that the two hemi-spheres $(x - y)\cdot\xi > 0$

and $(x-y)\cdot\xi<0$ make the same contribution. Thus

$$K(x,y)=\frac{1}{2(2\pi i)^n}\int\limits_{\substack{\Omega_\xi\\(x-y)\cdot\xi>0}}d\omega_\xi\oint\limits_{|\lambda|=M}\frac{e^{\lambda(x-y)\cdot\xi}}{P(\lambda\xi)}\lambda^{n-1}d\lambda.$$

We can put this expression into a form, which is invariant under affine transformations. We have for $n<m$

$$\text{(3.54b)}\qquad \oint\limits_{|\lambda|=M}\frac{e^{\lambda(x-y)\cdot\xi}}{P(\lambda\xi)}\lambda^{n-1}d\lambda=\oint\limits_{|\lambda|=M}\frac{e^{\lambda(x-y)\cdot\xi}-1}{P(\lambda\xi)}\lambda^{n-1}d\lambda$$

$$=\int\limits_{1/(x-y)\cdot\xi}^{\infty}r^{-2}dr\oint\limits_{|\lambda|=M}\frac{e^{\lambda/r}}{P(\lambda\xi)}\lambda^n d\lambda$$

$$=\int\limits_{1/(x-y)\cdot\xi}^{\infty}r^{n-1}dr\oint\limits_{|\lambda|=M/r}\frac{e^{\lambda}}{P(\lambda r\xi)}\lambda^n d\lambda.$$

Since here $1/r<|(x-y)\cdot\xi|\leqq|x-y|$, we can take the contour integral over a circle of radius $M|x-y|$. Put $z=r\xi$. Then $dz=r^{n-1}drd\omega_\xi$. In this way we obtain the representation

$$\text{(3.55)}\qquad K(x,y)=\frac{1}{2(2\pi i)^n}\int\limits_{(x-y)\cdot z>1}dz\oint\limits_{|\lambda|=M|x-y|}\frac{e^{\lambda}}{P(\lambda z)}\lambda^n d\lambda;$$

in which an integral over a half-space has taken the place of an integral over a hemi-sphere.

The formula becomes particularly simple for a homogeneous polynomial $P=Q$. Evaluating the contour integral we find

$$\text{(3.56)}\qquad K(x,y)=\frac{1}{2(2\pi i)^{n-1}(m-n-1)!}\int\limits_{(x-y)\cdot z>1}\frac{dz}{Q(z)}.$$

For $x-y=r\zeta$ this becomes

$$K(x,y)=\frac{1}{2(2\pi i)^{n-1}(m-n-1)!}\int\limits_{z\cdot\zeta>1/r}\frac{dz}{Q(z)}.$$

We differentiate both sides with respect to r. Since K is of degree $m - n$ in r, we have $dK/dr = (m - n)K/r$. Hence

$$(3.57) \qquad 2(2\pi i)^{n-1}(m-n)!K(x, y) = \frac{1}{r} \int\limits_{(x-y)\cdot z=1} \frac{dS_z}{Q(z)}.$$

Thus K is essentially the plane integral of the function $1/Q$. We can transform it into an integral over the surface $Q(\eta) = 1$. For this purpose we put $z = t\eta$, where $Q(\eta) = 1$, $t > 0$. Then

$$1 = (x - y) \cdot z = t(x - y) \cdot \eta, \; 1 = Q(\eta) = t^{-m}Q(z).$$

If dS_z and dS_η are surface elements corresponding to the same solid angle $d\omega$ from the origin, we have by (2.21a)

$$dS_z = |x - y| \; |z|^n \, d\omega, \; dS_\eta = \frac{1}{m} |\operatorname{grad} Q(\eta)| \; |\eta|^n d\omega.$$

Hence

$$\frac{dS_z}{|x-y|\,Q(z)} = \frac{mt^{n-m}}{|\operatorname{grad} Q(\eta)|} dS_\eta = \frac{m[(x-y)\cdot\eta]^{m-n}}{|\operatorname{grad} Q(\eta)|} dS_\eta.$$

It follows from (3.57) that

$$(3.58) \quad K(x, y) = \frac{m}{2(2\pi i)^{n-1}(m-n)!} \int\limits_{\substack{Q(\eta)=1 \\ (x-y)\cdot\eta>0}} \frac{[(x-y)\cdot\eta]^{m-n}}{|\operatorname{grad} Q(\eta)|} dS_\eta$$

$$= \frac{m}{4(2\pi i)^{n-1}(m-n)!} \int\limits_{Q(\eta)=1} \frac{|(x-y)\cdot\eta|^{m-n}}{|\operatorname{grad} Q(\eta)|} dS_\eta$$

since $m - n$ is an odd integer. Using the fact that $K(x, y)$ is a fundamental solution of the operator L we have from (3.2) for arbitrary $f(y)$ of class C_1 that vanish outside a bounded set

$$(3.59) \quad L\left(\int f(y)\left[\int\limits_{Q(\eta)=1} \frac{|(x-y)\cdot\eta|^{m-n}}{|\operatorname{grad} Q(\eta)|} dS_\eta\right] dy\right)$$

$$= \frac{4}{m}(2\pi i)^{n-1}(m-n)!f(x).$$

Identity (1.10) can be considered as the special case of (3.59) corresponding to $L = \Delta^{(n+k)/2}$. One can also transform the expression (3.58) for K into a volume integral over the interior of $Q(\eta) = 1$, using the homogeneous character of the integrand, and obtains

$$(3.60) \quad K(x, y) = \frac{m}{4(2\pi i)^{n-1}(m-n)!} \int_{Q(\eta) \leq 1} |(x-y) \cdot \eta|^{m-n} d\eta.$$

Let next n be an even number. If also P is homogeneous, $P = Q$, we have

$$v(x, \xi, p) = \frac{(x \cdot \xi - p)^m}{m! Q(\xi)}.$$

Then the regular part K_2 of K is given (see (3.47b)) by

$$(2\pi i)^n K_2(x, y) = -(\Delta_y)^{n/2} \int_{\Omega_\xi} \frac{[(x-y) \cdot \xi]^m}{m! Q(\xi)} d\omega_\xi \int_0^1 \frac{(1-t)^m - 1}{t} dt.$$

Since K_2 is a polynomial of degree $< m$, it is a solution of $L[K_2] = 0$, and can be neglected. This leaves by (3.47a)

$$(3.61) \quad m!(2\pi i)^n K(x, y) = -(\Delta_y)^{n/2} \int_{\Omega_\xi} \frac{[(x-y) \cdot \xi]^m \log |(x-y) \cdot \xi|}{Q(\xi)} d\omega_\xi.$$

For $m < n$ this yields (neglecting a regular solution)

$$(3.62) \quad K(x, y) = -(\Delta_y)^{(n-m)/2} \int_{\Omega_\xi} \frac{\log |(x-y) \cdot \xi|}{(2\pi i)^n Q(\xi)} d\omega_\xi.$$

If we put $x - y = r\zeta$ the coefficient of $\log r$ will vanish. K in this case is homogeneous in $x - y$ of degree $m - n$. If, on the other hand, $m \geqq n$, we get from (3.61) within a regular solution

$$(3.63) \quad K(x, y) = -\int_{\Omega_\xi} \frac{[(x-y) \cdot \xi]^{m-n} \log |(x-y) \cdot \xi|}{(m-n)! (2\pi i)^n Q(\xi)} d\omega_\xi. \text{ [39]}$$

[39] Formulae (3.63) and (3.54a) were given by Herglotz [2], p. 191, and related formulae by Bureau [3], p. 158. Fundamental solutions for equations with constant coefficients have been given earlier in the case of 2 dimensions by Somigliana [1], and of 3 dimensions by Fredholm [1].

This expression can be written in the form

$$(3.63\text{a}) \qquad (m-n)!\,(2\pi i)^n K(x,y) = -\, r^{m-n} \log r \int\limits_{\Omega_\xi} \frac{(\zeta\cdot\xi)^{m-n}}{Q(\xi)}\, d\omega_\xi$$

$$- \, r^{m-n} \int\limits_{\Omega_\xi} \frac{(\zeta\cdot\xi)^{m-n} \log|\zeta\cdot\xi|}{Q(\xi)}\, d\omega_\xi .$$

Here the factor of $\log r$ is a form of degree $m-n$ in $x-y$, and hence a regular solution of the differential equation, and the remaining term is homogeneous of degree $m-n$ in $x-y$.

Let more generally n be even and $< m$, and P not necessarily homogeneous. Then the singular part of K is, according to (3.47a), given by

$$K_1(x,y) = \frac{-1}{(2\pi i)^n} \int\limits_{\Omega_\xi} v^{(n)}(x,\xi,y\cdot\xi) \log|(x-y)\cdot\xi|\, d\omega_\xi .$$

(The terms arising from differentiating $\log|(x-y)\cdot\xi|$ with respect to y are all regular, and can be combined with K_2.) By (3.52) we have then

$$(3.64) \qquad (2\pi i)^{n+1} K_1(x,y)$$

$$= - \int\limits_{\Omega_\xi} \log|(x-y)\cdot\xi| \int\limits_{|\lambda|=M} \frac{e^{\lambda(x-y)\cdot\xi}}{P(\lambda\xi)} \lambda^{n-1}\, d\lambda\, d\omega_\xi .$$

Let $R(\xi)$ be any homogeneous function of degree -1, which is positive for $\xi \neq 0$ and even in ξ. For example $R(\xi) = Q^{-1/m}(\xi)$, where Q is the characteristic form of L. Replacing in (3.64) the term $\log|(x-y)\cdot\xi|$ by $\log|R(\xi)(x-y)\cdot\xi|$ only has the effect to change K_1 by a regular solution of the differential equation. Since the integral does not change in form if ξ and λ are replaced by $-\xi$ and $-\lambda$ the points on Ω_ξ with $(x-y)\cdot\xi > 0$ make the same contribution as those with $(x-y)\cdot\xi < 0$. (We make use here of the fact that $n-1$ is an odd integer!) We have then, since $n-1 < m-1$,

$$(2\pi i)^{n+1}K_1(x, y)$$

$$= -2 \int_{\substack{\Omega_\xi \\ (x-y)\cdot\xi>0}} \log\,[R(\xi)(x-y)\cdot\xi] \int_{|\lambda|=M} \frac{e^{\lambda(x-y)\cdot\xi}-1}{P(\lambda\xi)}\lambda^{n-1}\,d\lambda\,d\omega_\xi.$$

Now for $(x-y)\cdot\xi > 0$

$$\lambda \int_{1/(x-y)\cdot\xi}^{\infty} e^{\lambda/r}[\log\,(R(r\xi))]r^{-2}\,dr$$

$$= \lambda \int_{1/(x-y)\cdot\xi}^{\infty} r^{-2}e^{\lambda/r}(-\log r + \log R(\xi))\,dr$$

$$= (e^{\lambda(x-y)\cdot\xi}-1)\log\,[R(\xi)(x-y)\cdot\xi] - \int_{1/(x-y)\cdot\xi}^{\infty} \frac{e^{\lambda/r}-1}{r}\,dr$$

$$= (e^{\lambda(x-y)\cdot\xi}-1)\log\,[R(\xi)(x-y)\cdot\xi] - \int_0^1 \frac{e^{\lambda(x-y)\cdot\xi s}-1}{s}\,ds.$$

Hence

$$(2\pi i)^{n+1}K_1(x, y) = -2\int_{\Omega_\xi} d\omega_\xi \int_{1/(x-y)\cdot\xi}^{\infty} \frac{dr}{r^2} \oint_{|\lambda|=M} \frac{e^{\lambda/r}\log\,(R(r\xi))\,\lambda^n}{P(\lambda\xi)}\,d\lambda$$

$$+ 2\int_{\Omega_\xi} d\omega_\xi \int_0^1 \frac{ds}{s} \oint_{|\lambda|=M} \frac{e^{\lambda s(x-y)\cdot\xi}-1}{P(\lambda\xi)}\lambda^{n-1}\,d\lambda.$$

Here the second term is a regular analytic function of $x-y$, since the integrand is regular analytic in $x-y$, s, λ, ξ in the region of integration. Putting again $z = r\xi$, and replacing λ by λr in the contour integral, we find that the singular part of K is given by

$$K_1(x, y) = \frac{-2}{(2\pi i)^{n+1}} \int_{(x-y)\cdot z>1} \log R(z)\left[\oint_{|\lambda|=M|x-y|} \frac{\lambda^n e^\lambda}{P(\lambda z)}\,d\lambda\right]dz. \tag{3.65}$$

For P homogeneous $= Q$, this becomes

$$(3.66)\quad K_1(x, y) = \frac{-2}{(m-n-1)!\,(2\pi i)^n} \int_{|(x-y)|\cdot z>1} \frac{\log R(z)}{Q(z)}\,dz.$$

In this case K_1 is not only the singular part of a fundamental solution, but is itself a fundamental solution.

Fundamental solution of linear elliptic systems with analytic coefficients [40]

The construction given for the fundamental solution of a single linear elliptic equation can be used as well for systems of equations without essential changes. We consider a system of functions $u_t(x)$, where $t = 1, 2, \ldots, N$ and $x = (x_1, \ldots, x_n)$. There shall be given a set of linear differential operators L_{it} for $i, t = 1, \ldots, N$ with coefficients that are analytic functions of x. The system of equations considered shall be of the form

$$(3.67)\qquad L_{it}[u_t] = B_i(x), \quad (i,\ t = 1, \ldots, N),$$

where we use the common summation convention: A repeated index is to be summed over, unless a bar appears under the index. The aim is to construct a system of functions $K_t^{\alpha}(x, y)$ such that the symbolic equations

$$(3.68)\qquad L_{it}[K_t^{\tau}(x,\ y)] = \delta_i^{\tau}\delta(x-y)$$

are satisfied. Here δ_i^{τ} are the Kronecker deltas, whereas $\delta(x-y)$ denotes the Dirac function.

Let m_t be the order of the highest derivatives of u_t occurring in (3.67). We then consider all operators L_{it} as of order m_t, where for some L_{it} the coefficients of all highest order terms may vanish. Let $Q_{it}(x, \xi)$ be the characteristic form of L_{it}, which is of degree m_t in ξ, and possibly vanishes identically. The condition for ellipticity of the system is then that

$$(3.69)\qquad Q(x, \xi) = \text{determinant } (Q_{it}(x, \xi)) \neq 0 \text{ for } \xi \neq 0$$

[40] For construction of fundamental solutions of non-analytic linear elliptic systems by the parametrix method, see Lopatinskii [1], [2].

and all x in question.[40a] (This implies that $\Sigma_t m_t$ must be an even number for $n > 1$.)

If each $u_t(x)$ and all its derivatives of order $< m_t$ vanish on a hyper-plane $x \cdot \xi = p$, the equations (3.67) reduce to

$$(3.70) \qquad Q_{ij}(x, \xi)\left(\xi_\alpha \frac{\partial}{\partial x_\alpha}\right)^{m_j}[u_j(x)] = B_i(x)$$

along that plane. Condition (3.69) permits to solve these equations for each

$$\left(\xi_\alpha \frac{\partial}{\partial x_\alpha}\right)^{m_{\underline{j}}}[u_{\underline{j}}],$$

which we may call the *normal derivative* of u_t of order m_t. If $Q^{it}(x, \xi)$ denotes the elements of the reciprocal matrix to the matrix of the Q_{it}, we have for $x \cdot \xi = p$

$$(3.71) \qquad \left(\xi_\alpha \frac{\partial}{\partial x_\alpha}\right)^{m_{\underline{j}}} u_{\underline{j}}(x) = Q^{ji}(x, \xi)\, B_i(x).$$

The theory of Cauchy and Kowalewski applies here and shows that for analytic $B_i(x)$ we can, in the neighborhood of a point x of $x \cdot \xi = p$, find a solution system $u_j(x)$ of (3.67), for which the u_j and their derivatives of order $< m_j$ vanish on $x \cdot \xi = p$.

In particular we can find a system of functions $v_i^j(x, \xi, p)$, for which

$$(3.72) \qquad L_{it}[v_t^j(x, \xi, p)] = \delta_i^j$$

and for which the v_j^k and all their x-derivatives of order $< m_j$ vanish for $x \cdot \xi = p$. The $v_j^k(x, \xi, p)$ are homogeneous of degree 1 in ξ and p. If the coefficients of the L_{ik} are analytic in a neighborhood of the origin, the v_j^k wil be analytic in x, ξ, p in a set of the form (3.14) for some positive ε. In addition we can write

$$(3.73) \qquad v_i^j(x, \xi, p) = (x \cdot \xi - p)^{m_{\underline{i}}}\, w_{\underline{i}}^j(x, \xi, p),$$

[40a] This definition of ellipticity of a system agrees with that of Petrovskii [4], p. 16.

where the w_i^j are analytic in the set (3.14). Equations (3.72) show that

$$(3.74) \qquad m_t!\, Q_{it}(x, \xi)\, w_t^j(x, \xi, x \cdot \xi) = \delta_i^j$$

or that

$$(3.75) \qquad w_i^j(x, \xi, x \cdot \xi) = \frac{1}{m_{\underline{i}}!} Q^{\underline{i}j}(x, \xi),$$

where the bar under the letter i indicates that the index i is not to be summed over.

In analogy to (3.17) we form the system of functions

$$(3.76) \qquad V_j^i(x, \xi, p) = -\int_0^{x \cdot \xi - p} g(t) \frac{\partial}{\partial p} v_j^i(x, \xi, t + p)\, dt,$$

where $g(s)$ shall be the function defined by (3.19). Then

$$(3.77) \qquad L_{it}[V_t^j(x, \xi, p)] = \delta_i^j g(x \cdot \xi - p).$$

We then introduce

$$(3.78) \qquad W_j^i(x, y) = \int_{\Omega_\xi} V_j^i(x, \xi, y \cdot \xi)\, d\omega_\xi$$

and finally arrive at the fundamental solution matrix

$$(3.79) \qquad K_j^i(x, y) = (\Delta_y)^{(n+k)/2}\, W_j^i(x, y).$$

As before

$$W_j^i(x, y) = r^{m_{\underline{j}}+k}\left(A_{\underline{j}}^i(x, y, r, \zeta) + B_j^i(x, y, r, \zeta) \log r\right)$$

for $x - y = r\zeta$. Here the A_j^i and B_j^i are analytic in their arguments, and $B_j^i = 0$ for odd n. Correspondingly the K_j^i are of the form

$$K_j^i(x, y) = r^{m_{\underline{j}}-n}\left(A_{\underline{j}}''^i(x, y, r, \zeta) + B_{\underline{j}}''^i(x, y, r, \zeta) \log r\right).$$

The $K_j^i(x, y)$ form an analytic solution of

$$(3.80) \qquad L_{it}[K_t^j(x, y)] = 0$$

for $x \neq y$ and $|x - y|$ sufficiently small. More precisely we have for a system of functions $f_j(y)$ of class C_1

$$L_{it}\left[\int_D f_j(y) K_t^j(x, y)dy\right] = f_i(x). \tag{3.81}$$

The *adjoint* system of differential equations is given by

$$\bar{L}_{it}[u_i(x)] = 0, \tag{3.82}$$

where $\bar{L}_{it}$ is the adjoint operator of L_{it}, and thus of order m_t. Here a certain conceptual difficulty arises, if the m_t are not all equal. In that case the system (3.82) will no longer be elliptic in the sense defined here.[40b] We still have an identity corresponding to (3.36) of the type of Green

$$\int_R (u_i(x) L_{it}[v_t(x)] - v_t(x)\bar{L}_{it}[u_i(x)])\, dx = \int_B M_{it}[u_i, v_t]dS_x \tag{3.83}$$

where M_{it} is a bi-linear differential operator of order $m_t - 1$ in u_i and v_t. If we take for the $v_t(x)$ the functions $K_t^j(x, y)$ this reduces to

$$u_j(y) = \int_R K_t^j(x, y)\bar{L}_{it}[u_i(x)]dx + \int_B M_{it}[u_i(x), K_t^j(x, y)]dS_y. \tag{3.84}$$

It will be noted that the boundary integral contains in general derivatives of order $\operatorname{Max}_t m_t$ of each u_i. From this identity follows the analyticity of the solutions of the *adjoint* system (3.82). One is somewhat handicapped for unequal m_t to prove the analyticity of the solutions u_t of the original system (3.67) because of the difficulties of constructing a fundamental solution of the adjoint system. (A proof that does not make use of the adjoint will, however, be given in chapter VII.) If all m_t are equal, there is of course no difficulty.

Let in particular the operators L_{it} have constant coefficients.

[40b] This difficulty disappears with a more general definition of ellipticity suggested by L. Nirenberg. One assumes that the order of L_{it} does not necessarily depend on t alone, but is of the form $\mu_i + m_t$. Ellipticity is then still defined by definiteness of the form (3.69).

We can write

$$L_{it} = P_{it}\left(\frac{\partial}{\partial x}\right)$$

where P_{it} is a polynomial of degree $\leqq m_t$. Let P^{js} be the matrix reciprocal to that of the P_{it}. Then

$$v_t^j(x, \xi, p) = \frac{1}{2\pi i}\oint_C \frac{e^{\lambda(x\cdot\xi - p)}}{\lambda} P^{tj}(\lambda\xi)\, d\lambda$$

where C is a path in the complex λ-plane, which includes all roots of det $(P_{it}(\lambda\xi)) = 0$.

If all operators L_{it} are homogeneous, we have

$$P_{it} = Q_{it}$$

and

$$v_t^j(x, \xi, p) = \frac{1}{m_t!}(x\cdot\xi - p)^{m_t} P^{tj}(\xi). \tag{3.85}$$

For odd n we find then as in (3.44), (3.54)

$$K_t^j(x, y) = \frac{1}{4(2\pi i)^{n-1}(m_t - 1)!}(\Delta_y)^{(n-1)/2}\int_{\Omega_\xi}((x-y)\cdot\xi)^{m_t-1} \cdot \text{sign}\,[(x-y)\cdot\xi]P^{tj}(\xi)d\omega_\xi. \tag{3.86}$$

This expression is homogeneous of degree $m_t - n$ in $x - y$.

For even n we have, except for terms representing a regular solution of the differential equation, like in (3.61)

$$K_t^j(x, y) = \frac{-1}{m_t!(2\pi i)^n}(\Delta_y)^{n/2}\int_{\Omega_\xi} P^{tj}(\xi)\,((x-y)\cdot\xi)^{m_t} \cdot \log|(x-y)\cdot\xi|\, d\omega_\xi. \tag{3.87}$$

For $n > m_t$ the logarithmic part of K_t^j, i.e. the coefficient of $\log r$, vanishes.[41]

[41] Similar formulae have been given by Shapiro [1]. See also Bureau [9], ch. X, Morrey [2], [3] p. 104.

CHAPTER IV

Identities for Spherical Means

Symbolic expression for spherical means

Let $f(x)$ be a continuous function defined in a domain D. We put

$$I(x, r) = \frac{1}{\omega_n} \int_{\Omega_\xi} f(x + r\xi)\, d\omega_\xi. \tag{4.1}$$

Then $I(x, r)$ is defined for all x in D that have a distance $> r$ from the boundary of D. In what follows it will always be assumed that x and r satisfy this restriction, when $I(x, r)$ is used. From its definition $I(x, r)$ is an even function of r. It represents the average of f on a sphere of radius $|r|$ about the point x. The function I is continuous in x and r and satisfies

$$I(x, 0) = f(x). \tag{4.2}$$

More generally for f of class C_m the function $I(x, r)$ will also be of class C_m in x and r.

We can look at formula (4.1) for fixed r as defining a transformation T_r, which transforms a function $f(x)$ into a new function of x, which in addition depends on the parameter r:

$$I(x, r) = I(x, r; f) = T_r[f(x)]. \tag{4.2}$$

If we take here for f a function of the form $e^{i\eta \cdot x}$, where η is any constant vector, we find from (1.2), (1.5)

$$I(x, r) = T_r[e^{i\eta \cdot x}] = P_\nu(r\,|\,\eta\,|)e^{i\eta \cdot x}, \tag{4.3}$$

where $\nu = (n - 2)/2$, and $P_\nu(s)$ is a kind of "normalized" Bessel function

$$(4.4)\quad P_\nu(s) = 2^\nu\Gamma(\nu+1)s^{-\nu}J_\nu(s) = \frac{\Gamma(\nu+1)}{\sqrt{\pi}\,\Gamma(\nu+\frac{1}{2})}\int_{-1}^{1}(1-p^2)^{\nu-\frac{1}{2}}e^{isp}dp.$$ [42]

For the values of n in question 2ν is a non-negative integer, and $P_\nu(s)$ is an entire even function of s with $P_\nu(0) = 1$. Since for the Laplace operator Δ_x

$$(ir)^2\Delta_x[e^{i\eta\cdot x}] = (r\,|\,\eta\,|)^2\,e^{i\eta\cdot x},$$

we can write (4.3) in the form

$$(4.5)\qquad T_r[e^{i\eta\cdot x}] = P_\nu(ir\sqrt{\Delta_x})e^{i\eta\cdot x};$$

here the right hand side stands for what is obtained by expanding $P_\nu(ir\sqrt{\Delta_x})$ into a series of powers of Δ_x and having each term act separately on $e^{i\eta\cdot x}$. Since η is an arbitrary vector, formula (4.5) suggests the *symbolic* equation

$$(4.6)\qquad T_r = P_{(n-2)/2}(ir\sqrt{\Delta}).$$

Of course the interpretation of (4.6) by a power series expansion would not be legitimate for general continuous f.

The fundamental identity for iterated spherical means

For continuous $f(x)$ we define the *iterated spherical mean* $M(x, \lambda, \mu)$ by

$$(4.7)\qquad M(x,\lambda,\mu) = T_\lambda T_\mu[f(x)] = \frac{1}{\omega_n}\int_{\Omega_\zeta} I(x+\lambda\zeta,\mu)\,d\omega_\zeta.$$

If we substitute for I its expression from (4.1) we find

$$(4.8)\qquad M(x,\lambda,\mu) = \frac{1}{(\omega_n)^2}\int_{\Omega_\zeta}\int_{\Omega_\xi} f(x+\lambda\zeta+\mu\xi)\,d\omega_\xi\,d\omega_\zeta.$$

This shows that $M(x, \lambda, \mu)$ is symmetric in λ and μ

$$(4.9)\qquad M(x,\lambda,\mu) = M(x,\mu,\lambda).$$

[42] See Courant-Hilbert [1], vol. II, p. 261.

Moreover

(4.9a) $M(x, \lambda, 0) = M(x, 0, \lambda) = I(x, \lambda);\ M(x, 0, 0) = f(x).$

We shall derive now an identity that expresses $M(x, \lambda, \mu)$ by a single (one-dimensional) quadrature in terms of I. This identity can be based most easily on formula (1.2). [43] Let $g(s)$ be an

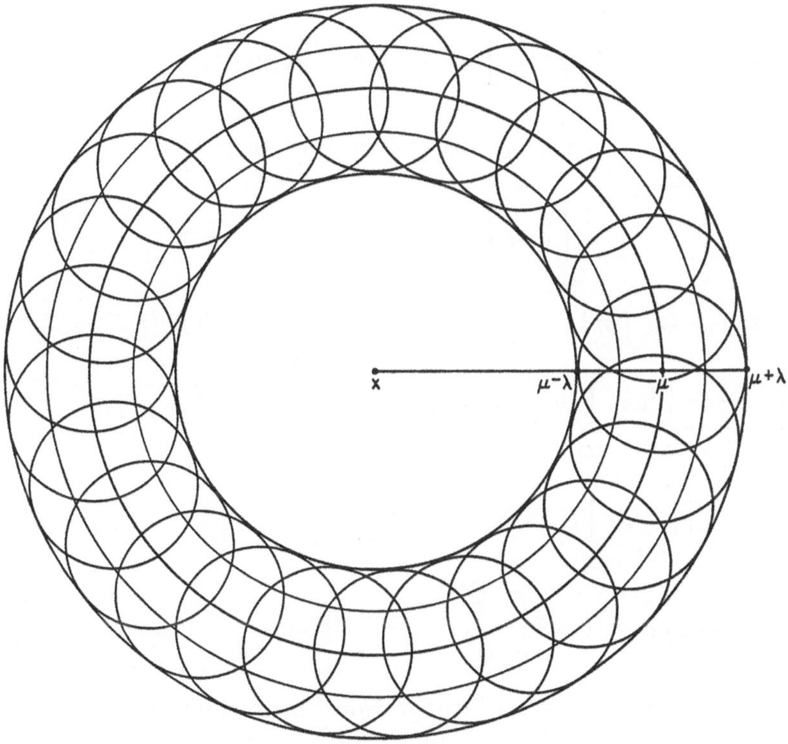

Figure 8

arbitrary continuous function, that vanishes outside a finite interval. We have

[43] A direct derivation is given in John [7], p. 171. The identity is a consequence of the fact that the shell $\mu - \lambda < |y - x| < \mu + \lambda$ in y-space can be swept out by spheres of radii r about x with $\mu - \lambda < r < \mu + \lambda$ as well as by spheres of radius λ with centers on the sphere $|y - x| = \mu$. (See Fig. 8.)

(4.9b) $$\omega_n^2 \int_0^\infty g(\lambda)\lambda^{n-1} M(x, \lambda, \mu) d\lambda = \int_0^\infty \lambda^{n-1} d\lambda \int_{\Omega_\zeta} d\omega_\zeta \int_{\Omega_\xi} d\omega_\xi f(x + \lambda\zeta + \mu\xi) g(\lambda).$$

Putting $\lambda\zeta = z$, the integral becomes

$$\int_{\Omega_\xi} d\omega_\xi \int dz\, f(x + z + \mu\xi) g(|z|).$$

Introducing $y = z + \mu\xi$ instead of z as variable of integration, we obtain

$$\int_{\Omega_\xi} d\omega_\xi \int dy\, f(x + y) g(|y - \mu\xi|).$$

For $y = r\eta$, $|\eta| = 1$, $r \geqq 0$ this becomes, using (1,2),

$$\int_0^\infty r^{n-1} dr \int_{\Omega_\eta} d\omega_\eta \int_{\Omega_\xi} d\omega_\xi f(x + r\eta) g(\sqrt{r^2 - 2r\mu\eta \cdot \xi + \mu^2})$$

$$= \omega_{n-1} \int_0^\infty r^{n-1} dr \int_{\Omega_\eta} f(x + r\eta) d\omega_\eta \int_{-1}^{+1} (1 - p^2)^{(n-3)/2} g(\sqrt{r^2 - 2r\mu p + \mu^2}) dp$$

$$= \omega_n \omega_{n-1} \int_0^\infty r^{n-1} I(x, r) \left[\int_{-1}^{+1} (1 - p^2)^{(n-3)/2} g(\sqrt{r^2 - 2r\mu p + \mu^2}) dp \right] dr.$$

Introducing a variable of integration λ instead of p by the relation $r^2 - 2r\mu p + \mu^2 = \lambda^2$, we obtain for $\mu > 0$

$$2\omega_n \omega_{n-1} \int_0^\infty r^{n-1} I(x, r) dr \int_{|\mu-r|}^{\mu+r} \frac{[(r+\mu-\lambda)(r+\mu+\lambda)(\lambda+r-\mu)(\lambda-r+\mu)]^{(n-3)/2}}{(2r\mu)^{n-2}} g(\lambda)\lambda d\lambda$$

$$= \frac{2\omega_n \omega_{n-1}}{(2\mu)^{n-2}} \int_0^\infty \lambda g(\lambda) d\lambda \int_{|\mu-\lambda|}^{\mu+\lambda} [(r+\mu-\lambda)(r+\mu+\lambda)(\lambda+r-\mu)(\lambda-r+\mu)]^{(n-3)/2} r I(x, r) dr.$$

Since the integrand is an odd function of r, we can replace $|\mu-\lambda|$ by $\mu - \lambda$ in the limit of integration. Comparing this expression with the original one (4.9b), from which it was derived, and using that $g(\lambda)$ was an *arbitrary* function, we find that for positive λ, μ

$$(4.9c)\quad M(x, \lambda, \mu)$$

$$= \frac{2\omega_{n-1}}{\omega_n(2\lambda\mu)^{n-2}} \int_{\mu-\lambda}^{\mu+\lambda} [(r+\mu-\lambda)(r+\mu+\lambda)(\lambda+r-\mu)(\lambda-r+\mu)]^{(n-3)/2} rI(x, r)dr,$$

which is the desired identity.[43a]

If we put $\beta = \mu + \lambda$, $\alpha = \mu - \lambda$, identity (4.9c) takes the form

$$(4.10)\quad M\left(x, \frac{\beta-\alpha}{2}, \frac{\beta+\alpha}{2}\right)$$

$$= \frac{\omega_{n-1}}{\omega_n} 2^{n-1}(\beta^2-\alpha^2)^{2-n} \int_\alpha^\beta [(\beta^2-r^2)(r^2-\alpha^2)]^{(n-3)/2} rI(x, r)dr$$

for $\beta > |\alpha|$.

In the special case, where $f(x) = e^{i\eta\cdot x}$ with $|\eta| = 1$, we have by (4.3)

$$I(x, r) = P_\nu(r)f(x), \quad M(x, \lambda, \mu) = P_\nu(\lambda)P_\nu(\mu)f(x).$$

Idendity (4.10) then goes over into

$$(4.11)\quad P_\nu\left(\frac{\beta-\alpha}{2}\right)P_\nu\left(\frac{\beta+\alpha}{2}\right)$$

$$= \frac{2^{2\nu+1}\Gamma(\nu+1)}{\sqrt{\pi}\Gamma(\nu+\frac{1}{2})}(\beta^2-\alpha^2)^{-2\nu}\int_\alpha^\beta [(\beta^2-r^2)(r^2-\alpha^2)]^{\nu-\frac{1}{2}} rP_\nu(r)dr,$$

where 2ν is assumed to be a non-negative integer. For $n = 3$ or $\nu = 1/2$ we have $P(s) = (\sin s)/s$, and (4.11) becomes one of the addition theorems of trigonometry.

If we express P_ν in terms of Bessel functions by (4.4), we obtain an identity for Bessel functions, which is essentially equivalent to (4.10):

[43a] This identity is given by Asgeirsson [1], p. 335 for the case of functions $f(x)$ depending on $|x|$ only.

$$(4.12)\quad J_\nu\left(\frac{\beta-\alpha}{2}\right)J_\nu\left(\frac{\beta+\alpha}{2}\right)$$
$$=\frac{2^{-\nu+1}(\beta^2-\alpha^2)^{-\nu}}{\sqrt{\pi}\,\Gamma(\nu+\frac{1}{2})}\int_\alpha^\beta[(\beta^2-r^2)(r^2-\alpha^2)]^{\nu-\frac{1}{2}}r^{1-\nu}J_\nu(r)\,dr.$$

Expression for a function in terms of its iterated spherical means

For the applications we need an expression for a function $f(x)$ in terms of its spherical means $I(x, r)$, where r is bounded away from 0. Such an expression will be obtained by first expressing $I(x, s)$ is terms of the iterated means $M(x, \lambda, \mu)$ with μ bounded away from 0, and then putting $s = 0$. We obtain an expression of the desired type by considering formula (4.10) for fixed β and variable α as an integral equation for $I(x, r)$, and by finding an explicit solution for this integral equation.

Equation (4.10) for fixed β can be reduced to an integral equation of the simpler type

$$(4.13)\qquad t^{n-2}h(t)=\frac{2\omega_{n-1}}{\omega_n}\int_0^t(t^2-p^2)^{(n-3)/2}g(p)\,dp$$

for a function g. [This equation is exactly equivalent to equation (1.2) in case of an even function $g(s)$. We could interpret $h(t)$ as the average on a sphere of radius t about the origin of a plane wave function $g(\xi\cdot x)$, where $|\xi| = 1$.] Equation (4.10) goes over into (4.13), if we make the substitutions

$$\beta^2-\alpha^2=t^2,\ \beta^2-r^2=p^2,\ g(p)=p^{n-2}I(x,\sqrt{\beta^2-p^2})$$

$$(4.14)\qquad h(t)=2^{2-n}t^{n-2}M\left(x,\frac{\beta-\sqrt{\beta^2-t^2}}{2},\frac{\beta+\sqrt{\beta^2-t^2}}{2}\right).$$

Assume that for a certain continuous $h(t)$ the integral equation (4.13) has a continuous solution $g(p)$ in an interval $0 \leqq t \leqq \beta$.

In order to obtain an expression for that solution we multiply equation (4.13) with $2t(s^2 - t^2)^{(n-3)/2}$, where s is a constant between 0 and β, and integrate with respect to t from 0 to s. Interchanging the integrations on the right hand side, and using that by (1.3)

$$2\int_p^s [(s^2 - t^2)(t^2 - p^2)]^{(n-3)/2} t\,dt = \left(\frac{s^2 - p^2}{2}\right)^{n-2} \int_{-1}^{+1} (1 - q^2)^{(n-3)/2} dq$$
$$= \frac{2^{2-n}\omega_n}{\omega_{n-1}} (s^2 - p^2)^{n-2},$$

we obtain

$$(4.15) \qquad 2^{n-2} \int_0^s t^{n-1}(s^2 - t^2)^{(n-3)/2} h(t)dt = \int_0^s (s^2 - p^2)^{n-2} g(p)dp.$$

The right hand side can certainly be differentiated $(n - 1)$-times with respect to s^2, even for $g(p)$ only continuous. We obtain in this way the expression

$$(4.16) \qquad g(s) = \frac{2^{n-1}s}{(n-2)!} \left(\frac{d}{ds^2}\right)^{n-1} \int_0^s t^{n-1}(s^2 - t^2)^{(n-3)/2} h(t)\, dt$$

whenever a continuous solution $g(s)$ exists. (Here d/ds^2 stands for $(1/2s)(d/ds)$.[44] In particular g is determined uniquely.

If n is an *odd* number $\geqq 3$, we can carry out all differentiations in (4.16) under the integral sign, and obtain

$$(4.17) \qquad g(s) = \frac{2^{n-2}\left(\dfrac{n-3}{2}\right)!}{(n-2)!}\, s \left(\frac{d}{ds^2}\right)^{(n-1)/2} s^{n-2} h(s).$$

This is an expression of the form

$$(4.18) \qquad g(s) = \sum_{k=0}^{(n-1)/2} c_{kn} s^k \frac{d^k h(s)}{ds^k}$$

[44] This derivation of the solution of (4.13) corresponds to the standard proceedure of solving integral equations of the type of Abel's equation or more generally of solving Volterra equations of the first kind. See Courant-Hilbert [1], vol. II, p. 414, and Volterra [1], ch. II.

with certain numerical constants c_{kn}. For example

$$g(s) = h(s) + sh'(s) \qquad \text{for } n = 3 \tag{4.19a}$$

$$g(s) = h(s) + \tfrac{5}{3} sh'(s) + \tfrac{1}{3} s^2 h''(s) \qquad \text{for } n = 5. \tag{4.19b}$$

If n is an *even* number $\geqq 2$, we can carry out only $(n - 2)/2$ of the differentiations under the integral sign in (4.16), and obtain

$$\begin{aligned} g(s) &= \frac{2s}{\left(\frac{n-2}{2}\right)!} \left(\frac{d}{ds^2}\right)^{n/2} \int_0^s \frac{t^{n-1}}{\sqrt{s^2 - t^2}} h(t)\,dt \\ &= \frac{2s}{\left(\frac{n-2}{2}\right)!} \left(\frac{d}{ds^2}\right)^{n/2} s^{n-1} \int_0^1 \frac{t^{n-1}}{\sqrt{1 - t^2}} h(st)\,dt. \end{aligned} \tag{4.20}$$

Carrying out the differentiations in this latter representation would involve h and its derivatives of order $\leqq n/2$.

We return now to the integral equation (4.10). Using the substitutions (4.14) we find from (4.17) for *odd* $n \geqq 3$ that

$$s^{n-2} I(x, \sqrt{\beta^2 - s^2}) = \frac{\left(\frac{n-3}{2}\right)!}{(n-2)!} s \left(\frac{d}{ds^2}\right)^{(n-1)/2} s^{2n-4} M\left(x, \frac{\beta - \sqrt{\beta^2 - s^2}}{2}, \frac{\beta + \sqrt{\beta^2 - s^2}}{2}\right).$$

In particular for $s = \beta$

$$f(x) = I(x, 0) = \frac{\left(\frac{n-3}{2}\right)!}{(n-2)!} \beta^{3-n} \left[\left(\frac{d}{ds^2}\right)^{(n-1)/2} s^{2n-4} M\left(x, \frac{\beta - \sqrt{\beta^2 - s^2}}{2}, \frac{\beta + \sqrt{\beta^2 - s^2}}{2}\right)\right]_{s=\beta}. \tag{4.21}$$

Here the derivative on the right is bound to exist, even if f, and hence I, are only continuous. Substituting $\beta^2 - s^2 = \alpha^2$ we obtain from (4.21)

$$(4.22)\quad f(x)=\frac{\left(\frac{n-3}{2}\right)!}{(n-2)!}\beta^{3-n}\left[\left(\frac{-d}{d\alpha^2}\right)^{(n-1)/2}(\beta^2-\alpha^2)^{n-2}M\left(x,\frac{\beta-\alpha}{2},\frac{\beta+\alpha}{2}\right)\right]_{\alpha=0}$$

where β is an arbitrary positive number. [45] We can simplify this formula under the assumption that $M(x, \lambda, \mu)$ is of class C_{n-1} in its arguments x, λ, μ for positive λ, μ. This assumption is certainly satisfied, if $f(x)$ is in C_{n-1}. However it holds even when it is only known that $I(x, r)$ is in C_{n-1} for $r > 0$. For M in C_{n-1} we have

$$(4.23)\quad M\left(x,\frac{\beta-\alpha}{2},\frac{\beta+\alpha}{2}\right) = \sum_{k=0}^{n-1}\frac{1}{k!}\alpha^k\left[\frac{\partial^k}{\partial\alpha^k}M\left(x,\frac{\beta-\alpha}{2},\frac{\beta+\alpha}{2}\right)\right]_{\alpha=0}+\varrho(\alpha),$$

where $\varrho(\alpha)$ is a function, whose $(n-1)$-st derivative with respect to α vanishes for $\alpha = 0$. Since from the symmetry of $M(x, \lambda, \mu)$ the left hand side of (4.23) is an even function of α, only even k contribute. (4.23) can be written

$$M\left(x,\frac{\beta-\alpha}{2},\frac{\beta+\alpha}{2}\right) = \sum_{\substack{k=0\\ k \text{ even}}}^{n-1}\frac{1}{k!\,2^k}\alpha^k\left[\left(\frac{\partial}{\partial\mu}-\frac{\partial}{\partial\lambda}\right)^k M(x,\lambda,\mu)\right]_{\lambda=\mu=\frac{\beta}{2}}+\varrho(\alpha).$$

Substituting this expression into (4.22) and evaluating the derivatives for $\alpha = 0$ we find

$$(4.24)\quad f(x)=\sum_{\substack{k=0\\ k \text{ even}}}^{n-1}\frac{(-1)^{k/2}\left(\frac{n-1}{2}\right)!\left(\frac{n-3}{2}\right)!}{k!\left(\frac{n-3+k}{2}\right)!\left(\frac{n-1-k}{2}\right)!}\lambda^k\left[\left(\frac{\partial}{\partial\mu}-\frac{\partial}{\partial\lambda}\right)^k M(x,\lambda,\mu)\right]_{\mu=\lambda}$$

where $\lambda = \beta/2$ is an arbitrary positive number. This expresses

[45] The expression (4.22) for odd n follows also directly from (4.10) by differentiation with respect to α^2. The detour by way of (4.15) is only essential for even n.

f in terms of M and its derivatives of order $\leqq n - 1$ for a positive pair of values λ, μ. In particular we have for $n = 3$

$$(4.25)\quad f(x) = M(x, \lambda, \lambda) - \frac{\lambda^2}{2}\left[M_{\lambda\lambda}(x, \lambda, \lambda) - 2M_{\lambda\mu}(x, \lambda, \lambda) + M_{\mu\mu}(x, \lambda, \lambda)\right].$$

If $f(x)$ is in C_2, this amounts to the identity for $n = 3$

$$f(x) = \frac{1}{16\pi^2}\int\limits_{\Omega_\xi}\int\limits_{\Omega_\eta}\Big[f(x + \lambda\xi + \lambda\eta) - \frac{\lambda^2}{2}\sum_{\substack{i,k=\\1,2,3}} f_{x_i x_k}(x + \lambda\xi + \lambda\eta)(\xi_i - \eta_i)(\xi_k - \eta_k)\Big]\,d\omega_\xi\, d\omega_\eta.$$

It is clear from (4.7) that if $I(x, r)$ is of class C_s in x, r for $r > 0$, then $M(x, \lambda, \mu)$ is of class C_s in x, λ, μ for $\mu > 0$. It follows from (4.24) for odd $n \geqq 3$, that $f(x)$ is at least of class C_{s-n+1}, if $I(x, r)$ is of class C_s for $r > 0$ (where $s \geqq n - 1$). Similarly the analyticity of $I(x, r)$ for $r > 0$ has as a consequence the analyticity of $f(x)$.

We turn now to the solution of (4.10) in the more complicated case of an *even* $n \geqq 2$. (For the applications to be discussed later one could get along with the simpler formulae derived for odd n, extending the results by Hadamard's method of descent to the case of an even number of dimensions.) We have from (4.20), (4.14)

$$s^{n-2} I(x, \sqrt{\beta^2 - s^2}) = \frac{2^{3-n}s}{\left(\frac{n-2}{2}\right)!}\left(\frac{d}{ds^2}\right)^{n/2}\int_0^s \frac{t^{2n-3}}{\sqrt{s^2 - t^2}} M\left(x, \frac{\beta - \sqrt{\beta^2 - t^2}}{2}, \frac{\beta + \sqrt{\beta^2 - t^2}}{2}\right) dt.$$

Putting $\beta^2 - s^2 = \alpha^2$, $\beta^2 - t^2 = p^2$, and finally setting $\alpha = 0$ we obtain

$$f(x) = \frac{(2\beta)^{3-n}}{\left(\frac{n-2}{2}\right)!}\left[\left(\frac{-d}{d\alpha^2}\right)^{n/2}\int_\alpha^\beta \frac{(\beta^2 - p^2)^{n-2}}{\sqrt{p^2 - \alpha^2}} M\left(x, \frac{\beta - p}{2}, \frac{\beta + p}{2}\right) p\, dp\right]_{\alpha=0}.$$

It is clear that this is already an expression for $f(x)$ in terms of the $M(x, \lambda, \mu)$ with $\mu \geqq \beta/2$, and hence in terms of $I(x, r)$ for $r \geqq \beta/2$. It remains to simplify the expression in such a way that it is possible to see the dependence of derivatives of f on derivatives of M.

The substitution

$$p^2 = \alpha^2 + (\beta^2 - \alpha^2)s^2$$

transforms the integral in the last formula into

$$(\beta^2 - \alpha^2)^{n-\frac{3}{2}} \int_0^1 (1 - s^2)^{n-2} M \, ds.$$

Observing that

$$\frac{dM}{d\alpha^2} = (1 - s^2) \frac{dM}{dp^2}$$

we can carry out the differentiations for $\alpha = 0$ and obtain

$$f(x) = \sum_{k=0}^{n/2} c_{kn} \beta^{2k} \int_0^1 (1 - s^2)^{n+k-2} \left[\left(\frac{d}{dp^2} \right)^k M \right]_{p=\beta s} ds \tag{4.26}$$

where the numerical factors c_{nk} are given by

$$c_{kn} = \frac{n 2^{2-n} (-1)^k \Gamma(n - \frac{1}{2})}{\left(\frac{n}{2} - k \right)! \, k! \, \Gamma\left(\frac{n-1}{2} + k \right)}. \tag{4.27}$$

For example we obtain for $n = 2$

$$f(x) = \int_0^1 \left(M - \frac{(1 - s^2)}{s} \frac{dM}{ds} \right) ds \tag{4.28}$$

where

$$M = M\left(x, \frac{\beta(1 - s)}{2}, \frac{\beta(1 + s)}{2} \right). \tag{4.29}$$

Let $I(x, r)$ be of class C_m in x and r for $r > 0$. Then $M(x, \lambda, \mu)$

is of class C_m for $\mu > 0$. Since $M(x, \lambda, \mu)$ is symmetric in λ and μ the function M given by (4.29) is even in s. Its derivatives of order $\leqq n/2$ with respect to s^2 depend continuously on derivatives of $M(x, \lambda, \mu)$ with respect to λ, μ of order $\leqq n$, and hence are of class C_{m-n} in x. It follows that for $I(x, r)$ of class C_m in x, r for $r > 0$, the function $f(x)$ is of class C_{m-n} in the case of even n. Similarly analyticity of $I(x, r)$ for $r > 0$ implies analyticity of $f(x)$.

The differential equation of Darboux

We have from (4.6) the symbolic equation

$$I(x, r) = T_r[f(x)] = P_\nu(ir\sqrt{\Delta})f(x). \tag{4.30}$$

Here the function

$$P_\nu(s) = 2^\nu \Gamma(\nu + 1) s^{-\nu} J_\nu(s); \quad \left[\nu = \frac{n-2}{2}\right], \tag{4.31}$$

is a solution of the differential equation

$$P_\nu''(s) + \frac{2\nu + 1}{s} P_\nu'(s) + P_\nu(s) = 0, \tag{4.32}$$

which follows from Bessel's differential equation

$$J_\nu''(s) + \frac{1}{s} J_\nu'(s) + \left(1 - \frac{\nu^2}{s^2}\right) J_\nu(s) = 0. \tag{4.33}$$

This suggests the partial differential equation

$$\left[\frac{2\nu + 1}{r}\frac{\partial}{\partial r} + \frac{\partial^2}{\partial r^2} - \Delta\right] P_\nu(ir\sqrt{\Delta}) = 0, \tag{4.34}$$

and makes it plausible that $I(x, r)$ is a solution of the Euler-Poisson-Darboux equation

$$\frac{\partial^2 I(x, r)}{\partial r^2} + \frac{n-1}{r}\frac{\partial I(x, r)}{\partial r} = \Delta_x I(x, r). \tag{4.35}$$

A direct proof of the equation (4.35) for the spherical means $I(x, r)$ of a function $f(x)$ is easily supplied, at least under the assumption that $f(x)$ is in C_2.[46] We have from the definition of $I(x, r)$

$$\omega_n \int_0^r \lambda^{n-1} I(x, \lambda) d\lambda = \int_{|y| \leq r} f(x + y) dy.$$

Applying the delta operator to both sides we find that

$$\begin{aligned} \omega_n \int_0^r \lambda^{n-1} \Delta_x I(x, \lambda) d\lambda &= \int_{|y| \leq r} \Delta_x f(x + y) dy \\ &= \frac{1}{r} \int_{|y|=r} \sum_{i=1}^n f_{x_i}(x + y) y_i dS = r^{n-1} \int_{\Omega_\eta} \sum_i f_{x_i}(x + r\eta) \eta_i d\omega_\eta \\ &= r^{n-1} \frac{\partial}{\partial r} \int_{\Omega_\eta} f(x + r\eta) d\omega_\eta = \omega_n r^{n-1} \frac{\partial}{\partial r} I(x, r). \end{aligned}$$

Differentiation with respect to r yields equation (4.35).

It follows from (4.7) for the iterated spherical mean $M(\lambda, \mu)$ of a function $f(x)$ in C_2 that

$$\Delta_x M(x, \lambda, \mu) = \frac{\partial^2}{\partial \lambda^2} M(x, \lambda, \mu) + \frac{n-1}{\lambda} \frac{\partial}{\partial \lambda} M(x, \lambda, \mu).$$

From the symmetry relation (4.9) for M we then have

$$\begin{aligned} \text{(4.36)} \qquad & \frac{\partial^2}{\partial \lambda^2} M(x, \lambda, \mu) + \frac{n-1}{\lambda} \frac{\partial}{\partial \lambda} M(x, \lambda, \mu) \\ &= \frac{\partial^2}{\partial \mu^2} M(x, \lambda, \mu) + \frac{n-1}{\mu} \frac{\partial}{\partial \mu} M(x, \lambda, \mu). \end{aligned}$$

[46] See Courant-Hilbert [1], vol. II, pp. 411—412; Asgeirsson [1], p. 327. For results on equation (4.35) without the restriction to positive integers ,n and for further references see the papers by Diaz and Weinberger [1], and A. Weinstein [1].

CHAPTER V

The Theorems of Asgeirsson and Howard [47]

Ellipsoidal means of a function

For the derivation of the theorems in this chapter it is convenient to introduce the integrals of a function $f(x)$ in n-space over a general ellipsoid

$$F(x) = \sum_{i,k=1}^{n} a_{ik}x_i x_k = r^2. \tag{5.1}$$

Here the a_{ik} are assumed to be the coefficients of a symmetric positive definite quadratic form. The matrix of the a_{ik} will be denoted by a, and its determinant by $|a|$.

Equation (5.1) can be written in matrix notation in the form

$$F(x) = x'ax = r^2 \tag{5.2}$$

where x is considered a column vector, and x' its adjoint.

There exists a non-singular matrix T of determinant $|T| \neq 0$ such that

$$T'aT = I \tag{5.3}$$

where I is the identity matrix. The linear substitution

$$x = Ty \tag{5.4}$$

will then transform $F(x)$ into $y'Iy = |y|^2$. If dS_x is the ordinary surface element of the surface $F(x) = \text{const.} = r^2$, we denote by $d\omega$ the solid angle from the origin in y-space formed by the points y corresponding to the points x of dS under the substitution (5.4). In other words

$$d\omega = r^{1-n}dS_y,$$

[47] See Asgeirsson [1], Courant-Hilbert [1], vol. II, pp. 417 et seq.; A. S. Howard [1].

where dS_y is the image of dS_x under (5.4). It is clear that $d\omega$ is determined by dS_x alone, independently of the choice of the matrix T, and is invariant under general affine transformations, since any linear transformation that carries spheres about the origin into such spheres preserves solid angles from the origin.

We can derive an expression for $d\omega$ in terms of dS_x and of the equation of the ellipsoid, which does not involve T. For that purpose we consider a volume integral

$$\int f(x)dx$$

extended over an arbitrary region in x-space. If we introduce the surfaces $F(x) = \text{const.} = r^2$ as coordinate surfaces, we can reduce the integral to a simple integral of integrals over those surfaces. One has then for the volume integral the expression [48]

$$\int_0^\infty 2r\,dr \int_{F(x)=r^2} \frac{f(x)dS_x}{|\text{grad } F(x)|}.$$

On the other hand the linear substitution (5.4) carries this volume integral into

$$|T| \int f(Ty)dy = |T| \int r^{n-1}dr \int_{|y|=r} f(Ty)d\omega.$$

Comparing, and observing that by (5.3)

$$|a|\ |T|^2 = 1,$$

we find the desired expression

$$d\omega = \frac{2r^{2-n}\sqrt{|a|}}{|\text{grad } F(x)|}dS_x. \tag{5.5}$$

This derivation also shows that for continuous $f(x)$

$$\int_{F(x)<R^2} f(x)dx = |a|^{-1/2} \int_0^R r^{n-1}dr \int_{F(x)=r^2} f(x)\,d\omega. \tag{5.6}$$

[48] See e.g. R. Courant, Differential and Integral Calculus, vol. II, pp. 301, 302.

The mean value theorem of Asgeirsson

We consider the n-dimensional space, where n is an even number: $n = 2m$, and write $y_1, \ldots, y_m$ for $x_1, \ldots, x_m$, and $z_1, \ldots, z_m$ for $x_{m+1}, \ldots, x_n$. Let $u(x) = u(y, z)$ be a function of class C_2. We form the expression

$$(5.7) \qquad I(\alpha, \beta) = \frac{1}{\omega_n} \int\limits_{F(x, \alpha, \beta)=1} u(y, z) d\omega$$

where

$$(5.8) \qquad F(x, \alpha, \beta) = \frac{1}{\alpha} \sum_{i=1}^{m} x_i^2 + \frac{1}{\beta} \sum_{i=m+1}^{n} x_i^2 = \frac{|y|^2}{\alpha} + \frac{|z|^2}{\beta}.$$

α and β are assumed to be positive constants. The integral in (5.7) is of dimension $n - 1 = 2m - 1$.

Using the invariance of $d\omega$ under affine transformations and applying the substitution

$$y = \sqrt{\alpha}\eta, \ z = \sqrt{\beta}\zeta,$$

we obtain from (5.7) the spherical integral

$$(5.8a) \qquad \mathrm{I}(\alpha, \beta) = \frac{1}{\omega_n} \int\limits_{|\eta|^2+|\zeta|^2=1} u(\sqrt{\alpha}\,\eta, \ \sqrt{\beta}\,\zeta)\, d\omega.$$

Then

$$\frac{\partial}{\partial\alpha} I(\alpha, \beta) = \frac{1}{2\omega_n\sqrt{\alpha}} \int\limits_{|\eta|^2+|\zeta|^2=1} \sum_{i=1}^{m} u_{y_i}(\sqrt{\alpha}\,\eta, \ \sqrt{\beta}\,\zeta)\eta_i\, d\omega$$

$$= \frac{1}{2\omega_n} \int\limits_{|\eta|^2+|\zeta|^2<1} \Delta_y u(\sqrt{\alpha}\,\eta, \ \sqrt{\beta}\,\zeta)\, d\eta\, d\zeta$$

$$= \frac{1}{2\omega_n} (\alpha\beta)^{-m/2} \int\limits_{F(x,\alpha,\beta)<1} \Delta_y u(y, z)\, dy\, dz.$$

A similar formula can be obtained for $\partial I/\partial\beta$. Combining the results we find

$$\frac{\partial}{\partial\alpha} I(\alpha, \beta) - \frac{\partial}{\partial\beta} I(\alpha, \beta) = \frac{1}{2\omega_n} (\alpha\beta)^{-m/2} \int\limits_{F(x,\alpha,\beta) \leqq 1} (\Delta_y - \Delta_z)\, u(y, z)\, dy\, dz.$$

In particular we find that for a solution $u(y, z)$ of the differential equation

$$(\Delta_y - \Delta_z)u = \sum_{i=1}^{m} u_{y_i y_i} - \sum_{i=1}^{m} u_{z_i z_i} = 0 \tag{5.9}$$

the identity

$$\left(\frac{\partial}{\partial\alpha} - \frac{\partial}{\partial\beta}\right) I(\alpha, \beta) = 0$$

holds, from which it follows that $I(\alpha, \beta)$ is a function of $\alpha + \beta$ alone, or that

$$I(\alpha + \lambda, \beta - \lambda) = I(\alpha, \beta) \text{ for } 0 < \lambda < \beta. \tag{5.10}$$

For $\alpha < \beta$ and $\lambda = \beta - \alpha$ we find

$$I(\alpha, \beta) = I(\beta, \alpha), \text{ for } \alpha, \beta > 0. \tag{5.11}$$

Now from (5.8a)

$$\begin{aligned}
\lim_{\beta\to+0} \omega_n I(\alpha, \beta) &= \int\limits_{|\eta|^2+|\zeta|^2=1} u(\sqrt{\alpha}\eta, 0)\, d\omega \\
&= \left[\frac{d}{dr} \int\limits_{|\eta|^2+|\zeta|^2\leqq r^2} u(\sqrt{\alpha}\eta, 0)\, d\eta\, d\zeta\right]_{r=1} \\
&= \left[\frac{d}{dr} \int\limits_{|\eta|^2\leqq r^2} u(\sqrt{\alpha}\eta, 0)\, d\eta \int\limits_{|\zeta|^2\leqq r^2-|\eta|^2} d\zeta\right]_{r=1} \\
&= \left[\frac{d}{dr}\frac{1}{m}\omega_m \int\limits_{|\eta|^2\leqq r^2} u(\sqrt{\alpha}\eta, 0)(r^2 - |\eta|^2)^{m/2} d\eta\right]_{r=1} \\
&= \omega_m \int\limits_{|\eta|^2<1} u(\sqrt{\alpha}\eta, 0)(1 - |\eta|^2)^{(m-2)/2}\, d\eta \\
&= \omega_m \alpha^{1-m} \int\limits_{|y|^2<\alpha} u(y, 0)(\alpha - |y|^2)^{(m-2)/2}\, dy.
\end{aligned}$$

An analogous formula shows that $\lim_{\alpha \to +0} I(\alpha, \beta) = I(0, \beta)$ exists and gives its value. It follows from the symmetry relation (5.11) for $\beta \to +0$ that

$$(5.12) \qquad I(\alpha,\ 0) = I(0,\ \alpha).$$

If we introduce the spherical averages over spheres in m-space

$$I_1(\lambda) = \frac{1}{\omega_m} \lambda^{1-m} \int_{|y|=\lambda} u(y, 0)\, dS_y$$

$$I_2(\lambda) = \frac{1}{\omega_m} \lambda^{1-m} \int_{|z|=\lambda} u(0, z)\, dS_z,$$

identity (5.12) for $\alpha = \lambda^2$ goes over into

$$\int_0^\lambda (\lambda^2 - p^2)^{(m-2)/2} p^{m-1} (I_1(p) - I_2(p))\, dp = 0.$$

This is an integral equation for $g(p) = p^{m-1}(I_1(p) - I_2(p))$ of the form (4.13) with $h = 0$. Since it was shown there that g is determined uniquely by h, it follows that

$$I_1(\lambda) = I_2(\lambda).$$

This is the theorem of Asgeirsson:

For a solution of the differential equation (5.9) *of class* C_2 *the average of* $u(y, z_0)$ *on a sphere of radius* λ *and center* y_0 *in* y*-space is the same as the average of* $u(y_0, z)$ *on a sphere of radius* λ *and center* z_0 *in* z*-space.*

(We have proved the theorem here only for $y_0 = z_0 = 0$. However, since it is obviously invariant under translation, it holds for general y_0, z_0.)

Applications to the equations of Darboux and the wave equation

The iterated spherical means $M(x, \lambda, \mu)$ of a function $f(x)$ of class C_2 satisfy the equation (4.36), which can be looked on as a special case of equation (5.9) in bi-polar coordinates. Consider more generally *any* function $M(\lambda, \mu)$ of class C_2 for $\lambda, \mu \geqq 0$ that satisfies the differential equation

$$(5.13) \qquad M_{\lambda\lambda} + \frac{n-1}{\lambda} M_\lambda = M_{\mu\mu} + \frac{n-1}{\mu} M_\mu \qquad (n \geqq 2).$$

This equation implies that $M_\lambda(0, \mu) = M_\mu(\lambda, 0) = 0$, and hence that $M(\lambda, \mu)$ can be continued into the whole plane as an even function of λ and μ. Let y and z now denote vectors in n-space. Then $u(y, z) = M(|y|, |z|)$ will be a function of class C_2 in y and z that satisfies the equation

$$\Delta_y u = \Delta_z u.$$

We can apply Asgeirsson's therorem. Take a point (y^0, z^0) in yz-space with $|y^0| = \alpha$, $|z^0| = \beta$, and spheres of radius r. Then the theorem gives

$$(5.14) \qquad \int_{\Omega_\eta} M(|y^0 + r\eta|, \beta)\, d\omega_\eta = \int_{\Omega_\zeta} M(\alpha, |z^0 + r\zeta|)\, d\omega_\zeta.$$

In the special case where $y^0 = z^0 = 0$ this yields

$$(5.15) \qquad M(r, 0) = M(0, r).$$

For general y^0, z^0 we have

$$|y^0 + r\eta| = \sqrt{\alpha^2 + 2ry^0 \cdot \eta + r^2},$$

so that $M(|y^0 + r\eta|, \beta)$ is a plane wave function of η. Applying identity (1.2), formula (5.14) goes over into

$$(5.16) \qquad \int_{-1}^{1} (1-p^2)^{(n-3)/2} M(\sqrt{\alpha^2 + 2\alpha r p + r^2}, \beta)\, dp$$
$$= \int_{-1}^{1} (1-p^2)^{(n-3)/2} M(\alpha, \sqrt{\beta^2 + 2r\beta p + r^2})\, dp.$$

For $\alpha = 0$ (5.16) becomes

$$(5.17) \quad M(r, \beta) = \frac{\omega_{n-1}}{\omega_n} \int_{-1}^{1} (1-p^2)^{(n-3)/2} M(0, \sqrt{\beta^2 + 2r\beta p + r^2})\, dp,$$

(which in the case that M is an iterated spherical mean, is equivalent to formula (4.9c)). A similar expression for $M(\alpha, r)$ is obtained from (5.16) for $\beta = 0$. Using (5.15) it follows that[49]

$$M(\lambda, \mu) = M(\mu, \lambda) \tag{5.18}$$

for every solution of (5.13) that is of class C_2 for $\lambda, \mu \geqq 0$.

We can obtain from (5.15) a uniqueness theorem for the initial value problem of the equation of Darboux (4.35).[50] Let $I(x, r)$ be a solution of (4.35) of class C_2 in x, r for $r \geqq 0$. Let $I(x, 0) = f(x)$. Introduce the spherical means of $I(x, r)$:

$$M(x, \lambda, \mu) = \frac{1}{\omega_n} \int_{\Omega_\zeta} I(x + \lambda\zeta, \mu) d\omega_\zeta.$$

Then M satisfies the equation of spherical means

$$M_{\lambda\lambda} + \frac{n-1}{\lambda} M_\lambda = \Delta_x M.$$

It also satisfies

$$M_{\mu\mu} + \frac{n-1}{\mu} M_\mu = \Delta_x M,$$

because I is a solution of (4.35). Hence $M(x, \lambda, \mu)$ is a solution of (5.13) of class C_2 for $\lambda, \mu \geqq 0$. Then equation (5.15) yields

$$\begin{aligned} M(x, r, 0) &= \frac{1}{\omega_n} \int_{\Omega_\zeta} f(x + \lambda\zeta) d\omega_\zeta \\ &= M(x, 0, r) = I(x, r). \end{aligned}$$

Thus $I(x, r)$ is determined uniquely as the spherical average of its initial values $f(x)$.

Following Asgeirsson we can use his theorem to derive the solution of the initial value problem for the wave equation

$$u_{tt}(y, t) = \Delta_y u(y, t). \tag{5.19}$$

This equation can be obtained from (5.9) by replacing m by n,

[49] Asgeirsson [1], p. 334.

[50] See also Courant-Hilbert [1], vol. II, p. 381; Asgeirsson [1], p. 330 for direct proofs.

and considering only solutions $u(y, z)$ that depend on $y_1, \ldots, y_n$, $z_1 = t$, but are independent of $z_2, \ldots, z_n$. Let $u(y, t)$ have the initial values

$$u(y, 0) = f(y), \quad u_t(y, 0) = 0, \tag{5.20}$$

so that $u(y, t)$ can be defined for negative t as an even function of t. We apply Asgeirsson's theorem to the case where $z^0 = 0$. Then

$$\frac{1}{\omega_n} \int_{\Omega_\eta} u(y^0 + r\eta, 0)\, d\omega_\eta = \frac{1}{\omega_n} \int_{\Omega_\zeta} u(y^0, r\zeta_1)\, d\omega_\zeta.$$

The left hand side of this equation is the spherical average of $f(y)$, which we denote by $I(y^0, r)$. The right hand side is the spherical integral of a plane wave function, which can be simplified by (1.2). Using the fact that $u(y, t)$ is even in t, we obtain

$$\begin{aligned} I(y^0, r) &= \frac{2\omega_{n-1}}{\omega_n} \int_0^1 (1 - p^2)^{(n-3)/2}\, u(y^0, rp)\,dp \\ &= \frac{2\omega_{n-1}}{\omega_n} r^{2-n} \int_0^r (r^2 - p^2)^{(n-3)/2} u(y^0, p)\, dp. \end{aligned}$$

This is an integral equation for $u(y^0, t)$ of the form (4.13) with $g(p) = u(y^0, p)$, $h(t) = I(y^0, t)$. Its solution is determined uniquely (for continuous u) and is given by (4.16). Hence

$$u(y^0, t) = \frac{2^{n-1} t}{(n-2)!} \left(\frac{d}{dt^2}\right)^{n-1} \int_0^t r^{n-1} (t^2 - r^2)^{(n-3)/2} I(y^0, r)\, dr. \tag{5.21}$$

The *symbolic* solution of the initial value problem (5.19), (5.20) is given by

$$u(y, t) = \cos\,(it\sqrt{\Delta_y})\, f(y). \tag{5.21a}$$

whereas by (4.6)

$$I(y, r) = P_\nu(ir\sqrt{\Delta_y}) f(y).$$

Thus (5.21) corresponds to the identity

$$(5.22)\quad \cos s = \frac{2^{n-1}s}{(n-2)!}\left(\frac{d}{ds^2}\right)^{n-1}\int_0^s r^{n-1}(s^2-r^2)^{(n-3)/2}P_\nu(r)\,dr$$

$$= \frac{2^{3\nu+1}\Gamma(\nu+1)\,s}{(2\nu)!}\left(\frac{d}{ds^2}\right)^{2\nu+1}\int_0^s r^{\nu+1}(s^2-r^2)^{\nu-\frac{1}{2}}J_\nu(r)\,dr,$$

when 2ν is a non-negative integer.

We can compare the form (5.21) of the solution of the wave equation with the formula (2.32). The t-derivative of the solution represented by (2.32) has the initial values (5.20). Hence we have also instead of (5.21) the alternative expression

$$(5.23)\quad u(y^0,t) = \frac{1}{(n-2)!}\left(\frac{d}{dt}\right)^{n-1}\int_0^t r(t^2-r^2)^{(n-3)/2}I(y^0,r)\,dr.$$

Obviously the two expressions on the right hand sides of (5.21) and (5.23) must be identical for all functions I. In addition to (5.22) we have then also the formula

$$(5.24)\quad \cos s = \frac{2^{\nu}\Gamma(\nu+1)}{(2\nu)!}\left(\frac{d}{ds}\right)^{2\nu+1}\int_0^s r^{1-\nu}(s^2-r^2)^{\nu-\frac{1}{2}}J_\nu(r)\,dr.$$

Formula (5.21) represents the solution of the initial value problem (5.19), (5.20), provided that problem has a solution of class C_2. The expression for u, when simplified, involves the derivatives of the initial function $f(x)$ of order $\leqq n/2$ for even n, of order $\leqq (n-1)/2$ for odd n. Sufficient for the existence of a solution u of class C_2 of the problem is that f is of class $C_{(n+4)/2}$ for even n, of class $C_{(n+3)/2}$ for odd n. [51] The solution can exist accidentally under less stringent assumptions on f, as in the case, when $f(x)$ depends on fewer than n arguments.

The addition theorem for the cosine leads to the *formal* identity

[51] See Diaz and Weinberger [1], p. 709.

$$2\cos(\lambda i\sqrt{\Delta})\cos(\mu i\sqrt{\Delta}) = \cos[(\lambda+\mu)i\sqrt{\Delta}] + \cos[(\lambda-\mu)i\sqrt{\Delta}].$$ [52]

For $\mu = \lambda$ we are thus formally led to the identity

$$f(x) = 2\cos^2(\lambda i\sqrt{\Delta})f - \cos(2\lambda i\sqrt{\Delta})f. \tag{5.25}$$

This identity also can be interpreted as expressing the function $f(x)$ in terms of the simple and of the iterated spherical means of f and of their derivatives of order $\leq n-1$, provided n is an odd number. Indeed for odd n the expression $\cos(\lambda i\sqrt{\Delta})f$ is by (5.21) a combination of the spherical means of f over spheres of radius λ, and of their derivatives of orders $\leq (n-1)/2$. Applying the operator twice gives an expression for f in terms of spherical means over spheres of radii λ and 2λ and derivatives of order $\leq n-1$. Use is made of the fact that n is odd, to keep the radii of the spheres involved away from 0. (This amounts to using Huygens' principle for the wave equation in the strong form.) We shall not make use of (5.25), since the justification of that identity for f, which are only continuous but have means $I(x, r)$ of class C_{n-1} for $r > 0$, appears difficult.

The identity of Aughtum S. Howard

The reasoning that had been employed here to derive Asgeirsson's theorem leads to a more general identity. Let $u(x) = u(x_1, \ldots, x_n)$ be a function of class C_2. We define for positive values of the α_i

$$\omega_n I(\alpha_1, \ldots, \alpha_n) = \int\limits_{\sum_i x_i^2/\alpha_i = 1} u(x)\, d\omega. \tag{5.26}$$

where $d\omega$ is again the affine invariant surface element of the ellipsoid. Then we have similarly as before for $x_i = \sqrt{\alpha_i}\,\xi_i$

$$\omega_n I(\alpha_1, \ldots, \alpha_n) = \int\limits_{|\xi|=1} u(\sqrt{\alpha_1}\,\xi_1, \ldots, \sqrt{\alpha_n}\,\xi_n)\, d\omega_\xi$$

[52] This identity suggests that the class of continuous functions $f(x)$, to which the operator $\cos(it\sqrt{\Delta})$, as defined by (5.21), is applicable once and yields a continuous function of x and t, is closed under this operation.

$$(5.27)\qquad \omega_n I_{\alpha_k} = \tfrac{1}{2}\alpha_k^{-1/2}\int\limits_{|\xi|=1} \xi_k u_{x_k}(\sqrt{\alpha_1}\,\xi_1, \ldots, \sqrt{\alpha_n}\,\xi_n)\,d\omega_\xi$$

$$= \tfrac{1}{2}\int\limits_{|\xi|<1} u_{x_k x_k}(\sqrt{\alpha_1}\,\xi_1, \ldots, \sqrt{\alpha_n}\,\xi_n)\,d\xi$$

$$= \tfrac{1}{2}(\alpha_1 \ldots \alpha_n)^{-1/2} \int\limits_{\sum x_i^2/\alpha_i<1} u_{x_k x_k}(x)\,dx.$$

Let more generally a and b be two symmetric matrices, where the matrix a belongs to a positive definite form. There exists then a non-singular matrix T such that

$$TaT' = I, \; TbT' = \beta$$

where I is the unit matrix, and β is a diagonal matrix

$$\beta = \begin{pmatrix} \beta_1 & 0 & \ldots & 0 \\ 0 & \beta_2 & \ldots & 0 \\ \cdot & \cdot & \cdot\;\cdot\;\cdot & \cdot \\ 0 & 0 & \ldots & \beta_n \end{pmatrix}.$$

Then

$$T(a + \lambda b)T' = I + \lambda\beta.$$

Assume that λ is restricted to values, for which

$$\alpha_k = 1 + \lambda\beta_k > 0 \text{ for } k = 1, \ldots, n,$$

so that the matrix $a + \lambda b$ is positive definite. We consider then for an arbitrary continuous function $f(y)$ the ellipsoidal means

$$J(\lambda) = \frac{1}{\omega_n} \int\limits_{y'(a+\lambda b)^{-1}y=1} f(y)\,d\omega.$$

Apply the substitution $y = T^{-1}x$ to the variables of integration. Then $f(y)$ goes over into the function $g(x) = f(T^{-1}x)$, and

$$J(\lambda) = \frac{1}{\omega_n} \int\limits_{x'(I+\lambda\beta)^{-1}x=1} g(x)\,d\omega$$

$$= \frac{1}{\omega_n} \int\limits_{\sum_i x_i^2/\alpha_i=1} g(x)\,d\omega = I(\alpha_1, \ldots, \alpha_n).$$

Then by (5.27)

$$\frac{dJ(\lambda)}{d\lambda} = \sum_k \beta_k I_{\alpha_k}(\alpha_1, \ldots, \alpha_n)$$
$$= \frac{1}{2\omega_n}(\alpha_1 \ldots \alpha_n)^{-1/2} \int_{\sum x_i^2/\alpha_i < 1} \sum_k \beta_k g_{x_k x_k}(x)\, dx.$$

Now

$$\alpha_1 \ldots \alpha_n = |I + \lambda\beta| = |T|^2 |a + \lambda b|$$
$$dx = |T|\, dy$$
$$\sum_k \beta_k \frac{\partial^2}{\partial x_k^2} = \left(\frac{\partial}{\partial x}\right)' \beta \left(\frac{\partial}{\partial x}\right) = \left(\frac{\partial}{\partial y}\right)' T^{-1}\beta T'^{-1}\left(\frac{\partial}{\partial y}\right) = \left(\frac{\partial}{\partial y}\right)' b \left(\frac{\partial}{\partial y}\right).$$

It follows that

$$\text{(5.28)} \qquad \frac{dJ(\lambda)}{d\lambda} = \frac{1}{2\omega_n}|a + \lambda b|^{-1/2} \int_{y'(a+\lambda b)^{-1}y<1} \left(\frac{\partial}{\partial y}\right)' b \left(\frac{\partial}{\partial y}\right) f(y)\, dy.$$

We introduce now the general ellipsoidical mean of a function $f(z) = f(z_1, \ldots, z_n)$. The ellipsoid shall be described by its center x, a matrix c and a scale factor μ. Its equation in running coordinates $z = x + y$ shall be

$$y'\left(\frac{c + c'}{2}\right)^{-1} y = \mu^2.$$

(Its equation in *tangential* coordinates is then

$$u'cu = \mu^{-2},$$

where u gives the coefficients in the equation $u \cdot y = 1$ of a tangent plane of the ellipsoid.) If $c + c'$ is a positive definite matrix and $\mu \neq 0$ we put

$$\text{(5.29)} \qquad I(x, c, \mu) = \frac{1}{\omega_n} \int_{y'\left(\frac{c+c'}{2}\right)^{-1} y = \mu^2} f(x + y)\, d\omega,$$

where $d\omega$ is the affine invariant surface element of the ellipsoid in y-space. When c is the unit matrix, I reduces to the spherical mean $I(x, \mu)$ of f. When c is a diagonal matrix with elements α_i in the diagonal, I goes over into the expression (5.26) formed for $f(x + y)$ as a function of y. Obviously I depends only on the symmetric part of c, and is homogeneous of degree 0 in c and μ^{-2}. We have indeed

$$(5.30) \qquad I(x, c, \mu) = I(x, \tfrac{1}{2}\mu^2 (c + c'), 1).$$

We want to find an expression for the derivatives of $I(x, c, \mu)$ with respect to the element c_{ik} of the matrix c. We have from (5.30)

$$\frac{\partial I(x, c, \mu)}{\partial c_{ik}} = \left(\frac{dI(x, a + \lambda b, 1)}{d\lambda}\right)_{\lambda=0},$$

where a is the positive definite symmetric matrix $\frac{1}{2}\mu^2(c + c')$, and b is the symmetric matrix with elements b_{rs} defined by

$$b_{rs} = \tfrac{1}{2}\mu^2(\delta_i^r \delta_k^s + \delta_i^s \delta_k^r)$$

in terms of Kronecker deltas.

We can compute the λ-derivative from formula (5.28). In our case

$$\left(\frac{\partial}{\partial y}\right)' b \left(\frac{\partial}{\partial y}\right) = \mu^2 \frac{\partial^2}{\partial y_i \partial y_k},$$

$$|a + \lambda b|_{\lambda=0} = |\tfrac{1}{2}\mu^2(c + c')| = \mu^{2n} \left|\frac{c + c'}{2}\right|.$$

We find then that

$$\frac{\partial I(x, c, \mu)}{\partial c_{ik}} = \frac{\mu^{2-n}}{2\omega_n} \left|\frac{c + c'}{2}\right|^{-1/2} \int\limits_{y'\left(\frac{c+c'}{2}\right)^{-1} y < \mu^2} f_{x_i x_k}(x + y) dy.$$

The volume integral occurring in this formula can be split up into ellipsoidal surface integrals with the help of formula (5.6):

$$\int\limits_{y'\left(\frac{c+c'}{2}\right)^{-1}y<\mu^2} f_{x_i x_k}(x+y)\,dy$$

$$= \left|\frac{c+c'}{2}\right|^{1/2} \int_0^\mu r^{n-1}\,dr \int\limits_{y'\left(\frac{c+c'}{2}\right)^{-1}y=r^2} f_{x_i x_k}(x+y)\,d\omega$$

$$= \omega_n \left|\frac{c+c'}{2}\right|^{1/2} \int_0^\mu r^{n-1} I_{x_i x_k}(x, c, r)\,dr.$$

In this way we obtain the *identity due to A. Howard.*

$$I_{c_{ik}}(x, c, \mu) = \frac{\mu^{2-n}}{2} \int_0^\mu r^{n-1} I_{x_i x_k}(x, c, r)\,dr. \tag{5.31}$$

A simple direct verification of (5.31) could be based on the Fourier transformation. Let $f(x)$ be a function of the form

$$f(x) = e^{i\xi\cdot x} = e^{i\xi' x},$$

where ξ' is the row vector adjoint to the column vector ξ. To compute

$$I(x, c, \mu) = \frac{1}{\omega_n} e^{i\xi\cdot x} \int\limits_{y'\left(\frac{c+c'}{2}\right)^{-1}y=\mu^2} e^{i\xi' y}\,d\omega$$

we transform the matrix $(c + c')/2$ by a non-singular T, so that

$$T\frac{c+c'}{2}T'$$

is the unit matrix. The for $z = Ty$

$$y'\left(\frac{c+c'}{2}\right)^{-1} y = |z|^2.$$

It follows from (4.3) that

$$I(x, c, \mu) = \frac{1}{\omega_n} e^{i\xi\cdot x} \int\limits_{|z|=\mu} e^{i\xi' T^{-1} z}\, d\omega$$

$$= e^{i\xi\cdot x} P_\nu(\mu \,|\, \xi' T^{-1} \,|) = e^{i\xi\cdot x} P_\nu(\mu\sqrt{\xi' c \xi}).$$

Equation (5.31) for the special f then reduces to the identity

$$P_\nu'(\lambda) = -\lambda^{1-n} \int\limits_0^\lambda s^{n-1} P_\nu(s) ds,$$

where $\lambda = \mu\sqrt{\xi' c\xi}$. This is however just another version of the differential equation (4.32) satisfied by P_ν. Since (5.31) is seen to be correct for all exponential functions f, it holds generally for all $f(x)$ of class C_2.

Applications of Howard's identity

Since $I(x, c, \mu)$ depends on c and μ only through the combination $\mu^2 c$, we have

(5.32)
$$I_\mu(x, c, \mu) = \frac{2}{\mu} \sum_{i,k} c_{ik} I_{c_{ik}}(x, c, \mu)$$

$$= \mu^{1-n} \int\limits_0^\mu r^{n-1} \sum_{i,k} c_{ik} I_{x_i x_k}(x, c, r) dr.$$

If c is the unit matrix, this gives

(5.33)
$$I_\mu(x, c, \mu) = \mu^{1-n} \int\limits_0^\mu r^{n-1} \Delta_x I(x, c, r) dr,$$

which is equivalent to Darboux' equation (4.35).

Let now $f(x)$ be a solution of the homogeneous second order differential equation

(5.34)
$$\sum_{i,k=1}^{n} a_{ik} f_{x_i x_k}(x) = 0$$

with constant a_{ik} forming a matrix a. Then of course also

(5.35)
$$\sum_{i,k=1}^{n} a_{ik} I_{x_i x_k}(x, c, \mu) = 0$$

for all μ and c in question. It follows from (5.31) that $I(x, c, \mu)$ as a function of c satisfies the first order equation

(5.36)
$$\sum_{i,k=1}^{n} a_{ik} I_{c_{ik}}(x, c, \mu) = 0.$$

This equation implies that

$$I(x, c + \lambda a, \mu)$$

is independent of λ. When a is the unit matrix, (5.34) becomes the Laplace equation, and the ellipsoids

$$y'(c + \lambda a)^{-1} y = \mu^2$$

are *confocal* for varying λ, if c is symmetric. We thus have the theorem that *the ellipsoidal mean of a potential function is constant for a family of confocal ellipsoids.* [53]

For more general matrices a we can formulate our result in an analogus manner: The ellipsoidical means of a solution of (5.34) are constant for a family of ellipsoids which are "confocal" with respect to a metric, in which the "absolute" has the tangential equation $\Sigma a_{ik} u_i u_k = 0$. If the family of ellipsoids with parameter λ contains degenerate elements, the ellipsoidal mean will reduce to an integral over a lower dimensional quadratic manifold. This is the case for the equation (5.9) considered by Asgeirsson, with respect to which the ellipsoids

$$\sum_{i=1}^{m} \frac{y_i^2}{a - \lambda} + \sum_{i=1}^{m} \frac{z_i^2}{\lambda} = 1$$

are "confocal." The ellipsoidal mean of u is the same for all λ. This family however contains two degenerate elements corresponding to $\lambda = 0$, a, which reduce to spheres in n-dimensional spaces (precisely speaking they are given by degenerate quadratic

[53] This theorem had been proved by Asgeirsson [1], p. 341 as an application of his identity. An analogous theorem for the wave equation is given by Serrin [1].

equations in *tangential coordinates*). The equality of the means for the two degenerate elements gives rise to Asgeirsson's theorem. Similarly for the potential equation one family of confocal surfaces is represented by a family of concentric spheres, which contains the center as degenerate element, and leads to Gauss' law of the arithmetic mean.

We can look more generally at identity (5.31) as a formula that transfroms second x-derivatives into first c-dericatives, and makes it possible to associate with a homogeneous differential equations in x with constant coefficients a different equation in c of half that order. We have for example, applying identity (5.31) twice,

$$(5.37)\quad I_{c_{ik}c_{rm}}(x, c, \mu) = \tfrac{1}{2}\mu^{2-n}\int_0^\mu r^{n-1} I_{x_i x_k c_{rm}}(x, c, r)dr$$

$$= \tfrac{1}{4}\mu^{2-n}\int_0^\mu r\,dr\int_0^r s^{n-1} I_{x_i x_k x_r x_m}(x, c, s)\,ds.$$

This formula shows that the value of the second derivative

$$(5.38)\qquad I_{c_{ik}c_{rm}}(x, c, \mu)$$

must be invariant under permutations of the indices i, k, r, m. If in addition $f(x)$ satisfies a fourth order differential equation with constant coefficients of the form

$$(5.39)\qquad \sum_{i,k,r,m} a_{ikrm} f_{x_i x_k x_r x_m}(x) = 0,$$

the function I will satisfy the second order equation

$$(5.40)\qquad \sum_{i,k,r,m} a_{ikrm} I_{c_{ik}c_{rm}}(x, c, \mu) = 0.$$

As an illustration consider the case, where $f(x)$ satisfies an equation

$$(5.41)\qquad \sum_{i,k} a_{ik} f_{x_i x_i x_k x_k}(x) = 0,$$

which involves only derivatives of even order with respect to each variable. We consider $I(x, c, \mu)$ for $x = 0$, $\mu = 1$ and for diagonal matrices c with diagonal elements $\gamma_1, \ldots, \gamma_n$. Then I reduces to a function $I(\gamma_1, \ldots, \gamma_n)$. This function will satisfy the equation

$$\sum_{i,k=1}^{n} a_{ik} I_{\gamma_i \gamma_k}(\gamma) = 0. \tag{5.42}$$

The mean value theorems available for the latter equation of second order will then give rise to identities involving the iterated ellipsoidal means of a solution f of (5.41). We have to observe however that c is restricted to matrices, for which $c + c'$ is positive definite, and hence $\gamma_i > 0$. One could consider, for instance, the case where the coefficients a_{ik} are such that the characteristic cone of (5.42) from the origin is contained in the first "octant". Then $I(\gamma)$ at the origin can be expressed in terms of values of $I(\gamma)$ and their first derivatives on an $(n - 1)$-dimensional manifold in the first octant. Since $I(0) = f(0)$ this leads to an expression for a solution f of (5.41) at the origin in terms of ellipsoidal integrals of f, where the ellipsoids keep a finite distance away from the origin.

Similar considerations apply to higher order equations.

CHAPTER VI

Determination of a Function from Its Integrals over Spheres of a Fixed Radius

Functions periodic in the mean

The problem to be discussed in this chapter consists in solving a special equation of the form

$$Tf = g, \tag{6.1}$$

where T is a linear operator converting functions $f(x) = f(x_1, \ldots, x_n)$ into functions $g(x)$, and where T is *invariant under translations*. This invariance of T means that, if T associates with a function $f(x)$ the function $g(x)$, then it maps $f(x+z)$ on $g(x+z)$ for any z.

Examples of equations of this type are represented by the equations [54]

$$g(x) = Tf = \int_R f(x+y)K(y)dy \tag{6.2}$$

where R is a fixed bounded region in y-space and $K(y)$ is continuous in R. Other examples would be presented by the operator T giving the solution of a hyperbolic equation with constant coefficients for fixed t in terms of the initial date, or by the operator giving the spherical mean of f over a sphere of radius r and center x for fixed r. [55] The simplest example is the equation

$$g(s) = \int_{-a}^{+a} f(s+t)dt \tag{6.3}$$

[54] See Schwartz [2], vol. 2, p. 64.

[55] Equations of the type (6.2) have been discussed by John [1] and Delsarte [1]; more general equations of type (6.1) by L. Schwartz [1].

for a function f of a single scalar variable s. The solution f of (6.3) is not unique. Indeed any continuous function $f(s)$ of period $2a$, for which

$$\int_{-a}^{a} f(s)ds = 0 \tag{6.4}$$

satisfies (6.3) with $g \equiv 0$. Delsarte has coined the term *periodic in the mean* ("moyenne-périodique") for any solution of (6.2) with $g \equiv 0$. This is suggested by the last example.

By decomposition into plane waves we can *formally* reduce equation (6.1) to an equation of the same type in one dimension. Let indeed $\Phi(s)$ be an arbitrary function of the scalar s. For any unit vector ξ we can form the plane wave function $f(x) = \Phi(\xi \cdot x)$. Then $f(x)$ has the property that $f(x + z) = f(x)$ for $z \cdot \xi = 0$. Because of the invariance of T under translations the image [56] $g(x) = Tf$ has then the same property: $g(x + z) = g(x)$ for $z \cdot \xi = 0$. Hence $g(x)$ is also a plane wave function, and depends only on $\xi \cdot x$, and of course on ξ: $g(x) = \gamma(x \cdot \xi, \xi)$. We have then

$$T\Phi(\xi \cdot x) = \gamma(x \cdot \xi, \xi).$$

For fixed ξ the expression $T\Phi(\xi \cdot x) = T_\xi \Phi(s)$ defines a linear operator acting on functions of a scalar s, and depending on a unit vector ξ as parameter. We have

$$T_\xi \Phi(s) = \gamma(s, \xi). \tag{6.5}$$

The operator T_ξ for fixed ξ is again invariant under translations:

$$\begin{aligned} T_\xi \Phi(s+t) &= T\Phi(\xi \cdot x + t) = T\Phi(\xi \cdot (x + t\xi)) = \gamma((x + t\xi) \cdot \xi, \xi) \\ &= \gamma(x \cdot \xi + t, \xi) = \gamma(s + t, \xi). \end{aligned}$$

In order to solve (6.1) for the case, where $g(x)$ is a plane wave function of the form $g(x) = \gamma(x \cdot \xi)$, it is sufficient to find a solution $\Phi(s)$ of the one-dimensional equation

$$T_\xi \Phi = \gamma. \tag{6.6}$$

[56] Assuming that T can be applied to f.

Then $f(x) = \Phi(x \cdot \xi)$ will be a solution of (6.1). Since every sufficiently regular $g(x)$ can be decomposed into plane waves by the formulae of Chapter I, we can expect to solve (6.1) by solving one-dimensional equations (6.6) depending on the parameter ξ. However, even if this scheme for solving (6.1) can be carried through and justified, there is no guarantee that it will furnish *all* solutions of (6.1).

In the case, where equation (6.1) is of the special form (6.2), we have

$$T_\xi \Phi = \int\limits_R \Phi((x+y) \cdot \xi)\, K(y)dy$$

$$= \int \Phi(x \cdot \xi + p) \int\limits_{\substack{y \text{ in } R \\ y \cdot \xi = p}} K(y)dS_y\, dp.$$

Hence equation (6.6) takes the form

$$\gamma(s) = T_\xi \Phi = \int \Phi(s+p) k(p, \xi)\, dp, \tag{6.7}$$

where the function $k(p, \xi)$ is defined for all p and all unit vectors ξ by

$$k(p, \xi) = \int\limits_{\substack{y \cdot \xi = p \\ y \text{ in } R}} K(y)dS_y. \tag{6.8}$$

Let here R be a convex body in y-space, which has the origin as an interior point. Let

$$\underset{y \text{ in } R}{\text{Maximum}}\ y \cdot \xi = b(\xi),\quad \underset{y \text{ in } R}{\text{Minimum}}\ y \cdot \xi = -a(\xi).\ [57]$$

Then $k(p, \xi) = 0$ unless $-a(\xi) \leqq p \leqq b(\xi)$. Hence (6.7) can be written in the form

$$\gamma(s) = \int_{-a(\xi)}^{b(\xi)} \Phi(s+p) k(p, \xi) dp. \tag{6.9}$$

[57] In a different notation $b(\xi) = H(\xi)$; $-a(\xi) = -b(-\xi) = -H(-\xi)$, where $H(\xi)$ is the "function of support" of the body R. See Bonnesen and Fenchel, Theorie der konvexen Körper (Springer, 1934), pp. 23—4.

Let $\Phi(s)$ be prescribed in the interval from $-a(\xi)$ to $b(\xi)$:

$$\Phi(s) = \psi(s) \text{ for } -a(\xi) < s < b(\xi). \tag{6.10}$$

Then we can write (6.9) in the form

$$\gamma(s) = \int_0^s \Phi(q+b(\xi))k(q+b(\xi)-s, \xi)\,dq + \int_{s-a(\xi)}^{b(\xi)} \psi(q)k(q-s, \xi)\,dq \tag{6.11}$$

for $0 < s < a(\xi) + b(\xi)$. This is an integral equation of the first kind of the type of Volterra for $\Phi(t)$ in the interval

$$b(\xi) < t < a(\xi) + 2b(\xi). \tag{6.12}$$

If this equation can be solved for $\Phi(t)$ in the interval (6.12), we get a similar integral equation for $\Phi(t)$ for the next interval of length $a + b$

$$a(\xi) + 2b(\xi) < t < 2a(\xi) + 3b(\xi).$$

Continuing in this manner and applying the analogous process for negative t, we obtain a sequence of integral equations similar to (6.11) determining $\Phi(t)$ for all t. The solvability of a Volterra equation of the first kind, depends essentially on the behavior of the kernel at the upper limit, i.e. on the behavior of $k(p, \xi)$ for p near $b(\xi)$. For a region R with a sufficiently regular boundary one can show that near $t = b(\xi)$ the function $k(t, \xi)$ behaves like

$$\frac{\omega_{n-1}}{n-1}\, 2^{(n-1)/2}(b(\xi) - p)^{(n-1)/2} \frac{K(y^0)}{\sqrt{\varrho(y^0)}},$$

where y^0 is the point of contact of the plane $y \cdot \xi = b(\xi)$ with the boundary of R and where $\varrho(y^0)$ is the product of the principal curvatures of the boundary surface at y^0. This expression shows that for $K(y) \neq 0$ on the boundary of R the integral equation (6.11) can be reduced by fractional differentiation of a suitable order with respect to s to a Volterra equation of the second kind, which in turn can be solved by iteration.

This suggests the possibility of solving (6.6) for Φ, where in addition the values of Φ are prescribed in the interval (6.10). In particular for $\gamma = 0$ there should exist plane wave solutions with prescribed values in (6.10). These solutions can be considered as functions, which are periodic in the mean.

More generally we can expect that there exists a solution $f(x)$ of (6.2) for $g \equiv 0$, where $f(x)$ agrees with a prescribed function $h(x)$ for x in R. $h(x)$ will have to satisfy certain conditions on the boundary of R, if f is to be regular. Thus if $f(x)$ is to be integrable we have the necessary condition

$$\int_R h(y)K(y)dy = 0,$$

If f is to have integrable first derivatives we must also have

$$\int_R h_{y_i}(y)K(y)dy = 0 \quad \text{for } i = 1, \ldots, n.$$

For an odd number of dimensions, we can write $h(x)$ as a combination of plane waves by (1.14):

$$h(x) = \int_{\Omega_\xi} \psi(x \cdot \xi, \xi)d\omega_\xi,$$

provided $h(x)$ is sufficiently regular and vanishes outside some bounded set. Here

$$\psi(p, \xi) = \frac{\partial^{n-1}}{\partial p^{n-1}} \frac{1}{2(2\pi i)^{n-1}} \int_{y \cdot \xi = p} h(y)\, dS_y.$$

Assuming that we can find a solution of

$$T_\xi \Phi(s, \xi) = 0$$

which is continuous in ξ and p, and for which

$$\Phi(s, \xi) = \psi(s, \xi) \text{ for } -a(\xi) < s < b(\xi),$$

we have in

$$f(x) = \int_{\Omega_\xi} \Phi(x \cdot \xi, \xi)d\omega_\xi$$

a mean periodic function, which agrees with $h(x)$ for x in R. Indeed $Tf = 0$, since $T\Phi(x \cdot \xi, \xi) = T_\xi \Phi(s, \xi) = 0$ for every ξ. Moreover for x in R the expression $\Phi(x \cdot \xi, \xi)$ involves only values of $\Phi(p, \xi)$ for which $-a(\xi) < p < b(\xi)$, and hence for which Φ can be replaced by ψ, so that f reduces to h.

Functions determined by their integrals over spheres of radius 1

We shall now specialize our operator T so that equation (6.1) is the equation

$$g(x) = \frac{1}{\omega_n} \int_{\Omega_y} f(x + y)\, d\omega_y, \tag{6.13}$$

and we shall also restrict the dimension n of our space to the value 3. We shall verify rigorously for this case some of the results obtained above on a heuristic basis. If we introduce the general spherical mean $I(x, r)$ and the iterated spherical mean $M(x, \lambda, \mu)$ of f as defined by (4.1), (4.7), equation (6.13) can be written

$$I(x, 1) = g(x). \tag{6.14}$$

We first consider the homogeneous equation

$$I(x, 1) = 0, \tag{6.15}$$

assuming f to be continuous for all x. (6.15) has as a consequence that

$$M(x, \lambda, 1) = 0 \tag{6.16}$$

for all λ. Using identity (4.9c) with $n = 3$, $\mu = 1$ it follows that

$$\int_{\lambda-1}^{\lambda+1} rI(x, r)\, dr = 0.$$

Differentiating with respect to λ we find that $rI(x, r)$ has the period 2. In particular from (6.15)

$$I(x, 2n + 1) = 0 \text{ for all integers } n.$$

We have as an immediate consequence a uniqueness theorem: *If* $f(x)$ *is a continuous solution of* (6.15) *and if*

$$\lim_{x\to\infty} |x| f(x) = 0, \tag{6.17}$$

then $f(x) \equiv 0$. For (6.17) implies that for any x

$$\lim_{r\to\infty} rI(x, r) = 0,$$

and hence, since $rI(x, r)$ is periodic, that $rI(x, r) \equiv 0$. Then also $I(x, r) \equiv 0$, and in particular $f(x) = I(x, 0) = 0$.

Instead of prescribing conditions at ∞ for a solution $f(x)$ of (6.15) we can also try to determine f uniquely by prescribing its values in a sphere about the origin. We have here the following theorem:

Let $f(x)$ *be a continuous solution of* (6.14). *Let* ε *be an arbitrary positive number. Then* $f(x)$ *is determined uniquely by its values in the sphere* $|x| < 1 + \varepsilon$. *On the other hand* $f(x)$ *is not determined by its values in the sphere* $|x| \leqq 1$, *even if we restrict* $f(x)$ *to functions of any class* C_m. [*It is determined, if we assume* f *to be in* C_∞.]

For the proof of the first part of this theorem we assume that $f(x)$ is a continuous solution of (6.15) with $f(x) = 0$ for $|x| < 1 + \varepsilon$. Then also obviously

$$I(x, r) = 0 \text{ for } |x| < \varepsilon,\ |r| \leqq 1.$$

Since $rI(x, r)$ as a function of r has the period 2, it follows that

$$I(x, r) = 0 \text{ for } |x| < \varepsilon \text{ and all } r.$$

We can conclude that $f(x)$ vanishes identically for all x. [58] We have the identity

$$\int_{\Omega_\xi} f(x + R\xi)(x_i + R\xi_i)\,d\omega_\xi = x_i \int_{\Omega_\xi} f(x + R\xi)\,d\omega_\xi + \frac{1}{R}\frac{\partial}{\partial x_i}\int_0^R r^2\,dr \int_{\Omega_\xi} f(x + r\xi)\,d\omega_\xi, \tag{6.18}$$

[58] See Courant-Hilbert [1], vol. II, p. 246, for this argument.

which follows immediately by writing the last integral as a volume integral and by carrying out the differentiation. Now (6.18) has as a consequence: If $f(x)$ is a continuous function with the property that the integral of f over any sphere with center in an ε-neighborhood of the origin vanishes, then $x_i f(x)$ has the same property. By repeated application of this argument we find that for any polynomial $P(x)$ the function $P(x)f(x)$ has the property that its integral over any sphere vanishes, whose center lies in an ε-neighborhood of the origin. Since the polynomials form a complete set of functions on any sphere, it follows that $f(x)$ vanishes on every sphere about the origin. Thus $f(x) \equiv 0$.

The same result can be obtained, if we assume that f is a solution of (6.15) of class C_∞, which vanishes for $|x| \leqq 1$. In that case we get from the periodicity of $rI(x, r)$ that $I(x, r)$ and all its derivatives exist and vanish for $x = 0$ and all r. It follows then from (6.16) that the spherical means of $x_i f(x)$ have the same property. Then the integral of $P(x)f(x)$ over any sphere about the origin vanishes, and hence f vanishes.

On the other hand we can construct for any finite m a non-trivial solution $f(x)$ of (6.15) of class C_m, which vanishes for $|x| \leqq 1$.[59] Let $\Phi(s)$ be any continuous function of period 2, for which

$$\int_{-1}^{+1} \Phi(s)ds = 0. \tag{6.19}$$

Then $\Phi(x \cdot \xi)$ is a solution of (6.15) for any ξ with $|\xi| = 1$. We can find a function $\Phi(s)$ which is of class C_m for all s, satisfies (6.19), and reduces to a polynomial in s for $|s| \leqq 1$. Such a function is obtained, e.g. by defining

$$\Phi(s) = s(1 - s^2)^{m+1} \text{ for } |s| \leqq 1 \tag{6.20}$$

and by continuing $\Phi(s)$ periodically for $|s| > 1$. We then define $f(x)$ by

$$f(x) = \int_{\Omega_\xi} \Phi(x \cdot \xi)Q(\xi)d\omega_\xi \tag{6.21}$$

[59] For the analogous, but more complicated, construction in the case of two dimensions see John [2], pp. 555 et seq.

where $Q(\xi)$ is a suitably chosen continuous function. It is clear that the f so defined is a solution of (6.15) of class C_m. Let μ be the degree of the polynomial to which $\Phi(s)$ reduces for $|s| \leqq 1$. (If (6.20) is used, we would have $\mu = 2m + 3$.) We choose for $Q(\xi)$ a function, which is orthogonal to all polynomials $P(\xi)$ of degree $\leqq \mu$ on the sphere $|\xi| = 1$, but is not orthogonal to the function $\Phi(2\xi_1)$ on that sphere. That is possible, since the polynomiale $P(\xi)$ of degree $\leqq \mu$ span a *finite dimensional* space, and the function $\Phi(2\xi_1)$ is certainly not a polynomial on that sphere. (In fact there is always some polynomial, which we could take for Q.) Then $f(x) = 0$ for $|x| \leqq 1$, since $|x \cdot \xi| \leqq 1$, and hence $\Phi(x \cdot \xi)$ reduces to a polynomial of degree μ in ξ on Ω_ξ. Moreover the function f constructed does not vanish identically, since for $x = (2, 0, 0)$

$$f(x) = \int_{\Omega_\xi} \Phi(2\xi_1)Q(\xi)d\omega_\xi \neq 0. \tag{6.21a}$$

This completes the proof of the theorem.

The function $f(x)$ given by (6.21) cannot vanish identically in any sphere of radius > 1 about the origin, since otherwise it would vanish everywhere by the first uniqueness theorem. We have moreover for that function f

$$I(x, r) \not\equiv 0 \text{ for } |x| \leqq \varepsilon,\ |r - 1| \leqq \varepsilon \tag{6.22}$$

where ε is any positive number. Indeed the relation

$$I(x, r) = 0 \text{ for } |x| \leqq \varepsilon,\ |r - 1| \leqq \varepsilon$$

combined with the fact that

$$I(x, r) = 0 \text{ for } |x| \leqq \varepsilon,\ |r| \leqq 1 - \varepsilon,$$

(which follows from $f(x) = 0$ for $|x| \leqq 1$) would imply that

$$I(x, r) = 0 \text{ for } -1 + \varepsilon \leqq r \leqq 1 + \varepsilon,\ |x| \leqq \varepsilon.$$

Since $rI(x, r)$ has period 2 in r, it would follows that

$$I(x, r) = 0 \text{ for } |x| \leqq \varepsilon \text{ and all } r.$$

This means that f is a continuous function, for which the integrals over all spheres with centers in an ε-neighborhood of the origin vanish. As was shown before, it would then follow that $f(x) = 0$ for all x, contrary to (6.21a).

For further discussion of the equation (6.14) and construction of its solutions, we restrict ourselves to functions $f(x)$ of class C_2. Then $I(x, r)$ is also in C_2, and by (2.32)

$$u(x, t) = tI(x, t) \tag{6.23}$$

is *the* solution of the wave equation

$$u_{tt} = u_{x_1 x_1} + u_{x_2 x_2} + u_{x_3 x_3} \tag{6.24}$$

with initial values

$$u = 0, \ u_t = f(x) \ \text{for} \ t = 0. \tag{6.25}$$

Solving equation (6.14) amounts then to finding solutions of (6.24) with prescribed values on the hyper-planes, $t = 1$ and $t = 0$ where $u(x, t)$ is required to be of class C_2 for $0 \leqq t \leqq 1$.

We can reformulate our uniqueness theorem in terms of u. We have first that $u(x, t)$ is determined uniquely by its values on $t = 1$ and $t = 0$, if we require that

$$\lim_{x \to 0} |x| \, u_t(x, 0) = 0.$$

$u(x, t)$ is also determined uniquely by the values of $u(x, 1)$ and $u(x, 0)$ for all x, if in addition $u_t(x, 0)$ is prescribed for $|x| < 1 + \varepsilon$. The solution u of (6.24) is *not determined* by the values of $u(x, 1)$ and $u(x, 0)$ for all x and of $u_t(x\ 0)$ for $|x| \leqq 1$, unless u is also required to be of class C_∞.

Ordinarily solutions u of the wave equation are determined from given values of $u(x, 0)$ and $u_t(x, 0)$. One is tempted to replace these data by those obtained from observation of the quantity $u(x, t)$ itself (instead of its t-derivative) at two different t, say $t = 0$ and $t = 1$. It has been seen here that u is not determined by such data, unless conditions at infinity are also imposed. However $u(x, t)$ is determined uniquely for all *integers* t by the values of $u(x, 0)$ and $u(x, 1)$ (because $u(x, t)$ is of period 2 in t

for $u(x, 0) = u(x, 1) = 0$). For determining u at non-integral values of t one would need further data, e.g. the value of $u_t(x, 0)$ for $|x| < 1 + \varepsilon$.

Let $f(x)$ be the special function given by (6.21) and $u(x, t) = tI(x, t)$ be the corresponding solution of (6.24), (6.25). Then (see Fig. 9 and (6.22))

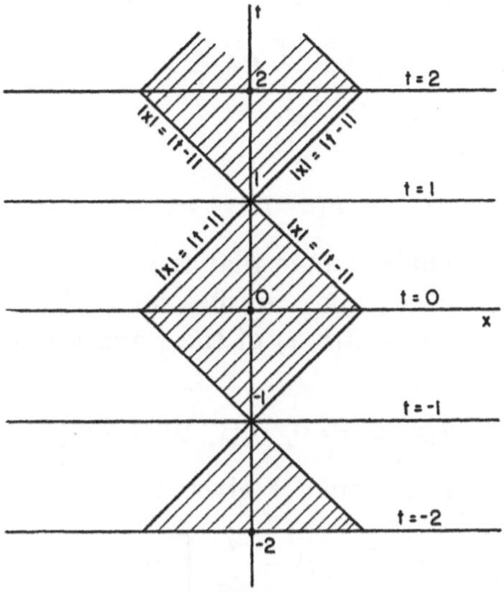

Figure 9

$u(x, t) = -u(x, -t) = u(x, 2+t) = -u(x, 2-t)$ for all x, t
$u(x, n) = 0$ for all integers n
$u(x, t) = 0$ for $|x| + |t| \leqq 1$.
$u(x, t) \not\equiv 0$ for (x, t) in a neighborhood of $x = 0$, $t = 1$.

It follows that

$$u(x, t) = 0 \text{ for } |x| = |t - 1|, \; 0 < t < 2.$$

We have as a consequence the theorem that the values of a solution $u(x, t)$ of (6.24) on a *finite* portion of a characteristic double cone $|x| = |1 - t|$ are not sufficient to determine the values of u in a full neighborhood of the vertex of the cone.

(This is proved here only for the portion of the double cone contained in $0 < t < 2$; it follows then for any bounded portion by applying a similarity transformation to x and t, which leaves equation (6.24) unchanged.) Hence the solution of this "exterior" characteristic boundary value problem has no finite domain of dependence.

We now turn to the solution of the homogeneous equation

$$I(x, 1) = \frac{1}{4\pi} \int_{\Omega_\xi} f(x + \xi)\, d\omega_\xi = 0 \tag{6.26}$$

where we prescribe

$$f(x) = h(x) \quad \text{for} \quad |x| \leqq 1. \tag{6.37}$$

For simplicity we shall assume that the prescribed function $h(x)$ is of class C_2 for all x, and vanishes for $|x| > 1$.

We form the solution $\bar{u}(x, t)$ of the wave equation with initial values

$$\bar{u}(x, 0) = 0, \; \bar{u}_t(x, 0) = h(x). \tag{6.28}$$

We know that this solution is given by

$$\bar{u}(x, t) = t\bar{I}(x, t) = \frac{t}{4\pi} \int_{\Omega_\xi} h(x + t\xi)\, d\omega_\xi. \tag{6.29}$$

Since $h(x)$ vanishes for $|x| \geqq 1$, we have

$$\bar{u}(x, t) = 0 \text{ unless } |x| - 1 < |t| < |x| + 1. \tag{6.30}$$

We then form the expression

$$u(x, t) = \sum_{n=-\infty}^{+\infty} \bar{u}(x, t + 2n). \tag{6.31}$$

Here the sum on the right hand side contains only a finite number of non-vanishing terms, since because of (6.30) only values of n with

$$|x| - 1 < |t + 2n| < |x| + 1 \tag{6.32}$$

contribute. Since $\bar{u}(x, t)$ is a solution of the wave equation of class C_2, the same holds then for $u(x, t)$. Moreover by definition (6.30) $u(x, t)$ has period 2 in t. Since $\bar{u}(x, t)$ is an odd function of t, the function $u(x, t)$ is also odd in t. As an odd function of period 2 the function $u(x, t)$ vanishes for all integers t. Putting

$$f(x) = u_t(x, 0), \; I(x, t) = \frac{1}{4\pi} \int_{\Omega_\xi} f(x + t\xi)\, d\omega_\xi,$$

we have by (6.23) $u(x, t) = tI(x, t)$, since $u(x, 0) = 0$. Since also $u(x, 1) = 0$, we see that $f(x)$ is a solution of (6.26) of class C_1. We find from (6.31), (6.30) that for $|x| < 1$

$$f(x) = \sum_{n=-\infty}^{+\infty} \bar{u}_t(x, 2n) = \bar{u}_t(x, 0) = h(x).$$

Thus the $f(x)$ constructed satisfies both conditions (6.26) and (6.27).

For $|x| > 1$ we can combine the terms with n and with $-n$, which make the same contribution. We find that then (see Fig. 10)

$$f(x) = 2\bar{u}_t(x, 2n) = \left[\frac{\partial}{\partial t} \frac{t}{2\pi} \int_{\Omega_\xi} h(x + t\xi) d\omega_\xi\right]_{t=2n} \tag{6.33}$$

where n is that positive integer for which

$$|x| - 1 < 2n \leqq |x| + 1. \tag{6.34}$$

In a similar fashion one can find a special solution of the *inhomogeneous* equation (6.13). We assume here that $g(x)$ is of class C_3 and that $g(x)$ and its derivatives of order $\leqq 3$ go to 0 at least like $|x|^{-3}$ for $x \to \infty$. We form the solution $\bar{u}(x, t)$ of the wave equation with initial values

$$\bar{u}(x, 0) = 0, \; \bar{u}_t(x, 0) = g(x).$$

We have then

$$\bar{u}(x, t) = t\bar{I}(x, t),$$

where $\bar{I}(x, t)$ denotes the spherical mean of g. We form the expression

$$u(x, t) = -\sum_{n=0}^{\infty} [\bar{u}_t(x, 2n + 1 + t) - \bar{u}_t(x, 2n + 1 - t)].$$

This is again a solution of the wave equation under the assumptions made for $g(x)$. Obviously $u(x, 0) = 0$. We put

$$f(x) = u_t(x, 0).$$

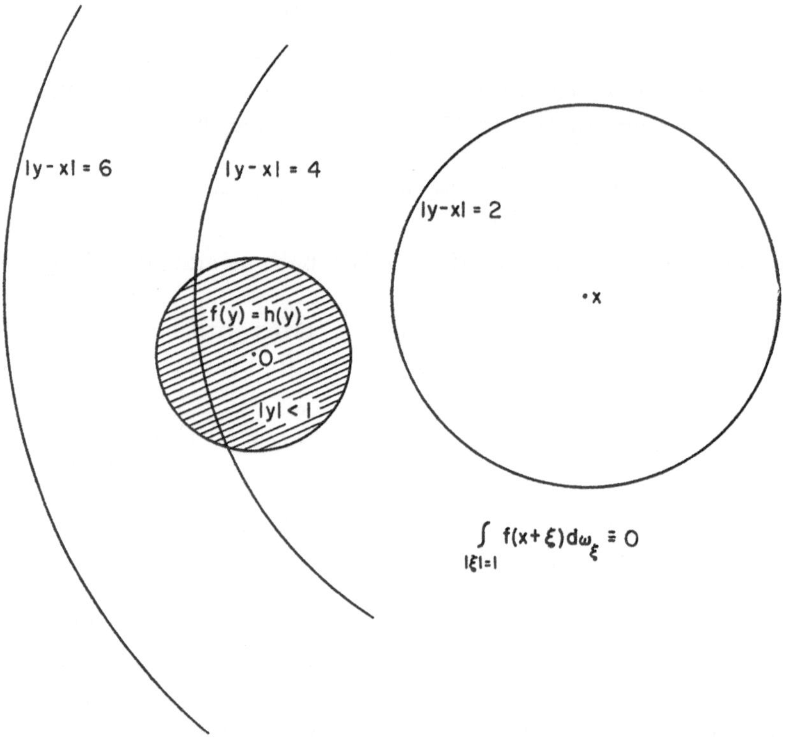

Figure 10

Then the average of $f(x)$ on a sphere of radius 1 and center x is given by

$$u(x, 1) = -\sum_{n=0}^{\infty} [\bar{u}_t(x, 2n + 2) - \bar{u}_t(x, 2n)] = \bar{u}_t(x, 0) = g(x).$$

Thus $f(x)$ is a solution of equation (6.13) of class C_1. Here $f(x)$

can be written

$$(6.35)\quad f(x) = -2\sum_{n=0}^{\infty} \bar{u}_{tt}(x, 2n+1) = -\Delta_x \sum_{n=0}^{\infty} 2\bar{u}(x, 2n+1)$$

$$= -2\Delta_x \sum_{n=0}^{\infty} (2n+1)\bar{I}(x, 2n+1)$$

$$= \sum_{n=0}^{\infty} \frac{-1}{2\pi(2n+1)} \int_{|y|=2n+1} \Delta_x g(x+y)\, dS_y.$$

Determination of a field of forces from its effect on a mobile sphere

As a physical application we consider the problem of determining a field of force by measuring its effect on a "homogeneous" sphere moved about in the field. We assume that the sphere does not affect the field, that the forces only act on the surface of the sphere, and that we can measure for every position of the sphere the resultant force and resultant moment. We take the sphere to have radius 1, and the force field to have components $F_i(x)$ for $i=1, 2, 3$ defined everywhere. (An example in question would be a gravitational field acting on a "thin" spherical shell of radius 1.) We ask ourselves how far the field can be determined in this way. If the force field has no effect on the body we have the equations

$$(6.36)\quad \int_{\Omega_\xi} F_i(x+\xi)\, d\omega_\xi = 0, \quad \int_{\Omega_\xi} [F_i(x+\xi)\xi_k - F_k(x+\xi)\xi_i]\, d\omega_\xi = 0$$

satisfied for all x and $i, k = 1, 2, 3$.

We denote by $I_i(x, r)$ the spherical average of the function $F_i(x)$ on a sphere of radius r about the point x, and by $I_{ik}(x, r)$ the spherical average of the function $F_i(x)x_k - F_k(x)x_i$. Then by (6.36) the $I_i(x, r)$ and $I_{ik}(x, r)$ vanish for $r = 1$. It follows as before that $rI_i(x, r)$ and $rI_{ik}(x, r)$ are periodic functions of r of period 2. We make use of the identity

$$(6.37)\quad \int_0^r \left[\frac{\partial I_i(x, \lambda)}{\partial x_k} - \frac{\partial I_k(x, \lambda)}{\partial x_i}\right] \lambda^2\, d\lambda$$

$$= r[I_{ik}(x, r) + x_i I_k(x, r) - x_k I_i(x, r)].$$

which is easily verified by writing the left hand side as a volume integral. The expression on the right of (6.37) is of period 2 in r. The same holds then for the expression on the left, and then also for its derivative with respect to r. Thus

$$r^2\left(\frac{\partial I_i(x, r)}{\partial x_k} - \frac{\partial I_k(x, r)}{\partial x_i}\right)$$

has period 2. Since also

$$r\left(\frac{\partial I_i(x, r)}{\partial x_k} - \frac{\partial I_k(x, r)}{\partial x_i}\right)$$

has period 2, it follows that

$$\frac{\partial I_i(x, r)}{\partial x_k} - \frac{\partial I_k(x, r)}{\partial x_i} \equiv 0.$$

In particular for $r = 0$

$$\frac{\partial F_i(x)}{\partial x_k} - \frac{\partial F_k(x)}{\partial x_i} \equiv 0.$$

Hence there exists a function $f(x)$ with

$$F_i(x) = \frac{\partial f(x)}{\partial x_i}.$$

We have from (6.36)

$$\frac{\partial}{\partial x_i} \int_{\Omega_\xi} f(x + \xi) d\omega_\xi = 0 \text{ for } i = 1, 2, 3.$$

Hence the spherical mean of $f(x)$ over spheres of radius 1 is independent of the center. Changing f by a suitable additive constant, we can bring about that

$$\int_{\Omega_\xi} f(x + \xi) d\omega_\xi = 0. \tag{6.38}$$

Thus the most general force field that will have no effect on a sphere of radius 1 must have a "potential" $f(x)$, which is periodic in the mean, i.e. satisfies (6.38). More precisely the potential $f(x)$ of the forces exists for $|x| < R$, if equations (6.36) hold for $|x| < R + 1$. A non-conservative force field will have an effect on the sphere for some positions of its center.

CHAPTER VII

Differentiability Properties of Solutions of Elliptic Systems

Canonical systems of differential equations

We consider a set of functions $u_1, \ldots, u_N$ of the independent variables $x = (x_1, \ldots, x_n)$. For simplicity we denote by $u_{i,\alpha}$ the partial derivative of u_i with respect to x_α. Generally suffixes following a comma denote differentiation with respect to the corresponding x-variable. Moreover the summation convention will be used again.

An *element* in x-space consists of a point and a plane passing through the point. It can be described by the coordinates x of the point and the direction numbers ξ of the normal of the plane, and will be denoted by (x, ξ). Thus for $\xi \neq 0$ the symbol (x, ξ) stands for the combination of the point x and of the plane $(y - x) \cdot \xi = 0$ in running coordinates y.

The u_i are said to satisfy a *canonical system of differential equations*, if for every x of a certain domain D there exist rational homogeneous functions $T_{m\alpha}$, $S_{mr\alpha\beta}$ of ξ of degree 0, which satisfy the identity

$$S_{mr\alpha\beta}\, \xi_\beta = 0, \tag{7.1}$$

and for which

$$u_{m,\alpha} = T_{m\alpha} + S_{mr\alpha\beta}\, u_{r,\beta}, \tag{7.2}$$

whenever x is in D and ξ is different from 0 and not a pole of any $T_{m\alpha}$ or $S_{mr\alpha\beta}$.

Considered as a system of differential equations (7.2) is highly *over-determined*, since we have such equations not only for every point x but for every element through the point. It will be seen however that well determined systems of differential equations yield canonical systems. The system (7.2), (7.1) expresses any first derivative of a u_m in terms of first derivatives of all of the u_k, which are in directions tangential to the element (x, ξ). If ξ is different from 0 and such that one of the rational functions $T_{m\alpha}$ or $S_{mr\alpha\beta}$ becomes infinite for that value of ξ (i.e. has ξ as a pole), we call the element (x, ξ) *characteristic* with respect to the system (7.2). An element (x, ξ), which is not characteristic will be called *free*. If all elements (x, ξ) are free, i.e. if the rational functions $T_{m\alpha}$ and $S_{mr\alpha\beta}$ have no real poles $\xi \neq 0$, we call the system of canonical equations *elliptic*. If the $T_{m\alpha}$, $S_{mr\alpha\beta}$ are prescribed functions of x, ξ and the u_i, we call the system (7.2) *quasi-linear*. (For any solution system $u_i(x)$ the T and S are then functions of x and ξ alone). It is assumed here that for the $u_i(x)$ admitted the $T_{m\alpha}$ and $S_{mr\alpha\beta}$ are for fixed x and hence fixed values of $u_i(x)$ rational homogeneous functions of degree 0 in ξ for all ξ, and satisfy (7.1). If the $S_{mr\alpha\beta}$ are independent of the u_i and the T_{mr} are linear in the u_i the canonical system is called *linear*.

Reduction of determined systems of differential equations to canonical form

We start with the decomposition of any first derivative $u_{,\alpha}$ of a function $u(x)$ into a "normal" and a "tangential" derivative with respect to an element (x, ξ). Denote by ϱ^2 the expression $\xi^2 = \xi \cdot \xi$, which is a quadratic form in ξ. Then we have identically in x and ξ

$$u_{,\alpha}(x) = \varrho^{-2}\xi_\alpha\xi_\beta u_{,\beta} + (\delta_\alpha^\beta - \varrho^{-2}\xi_\alpha\xi_\beta)u_{,\beta}. \tag{7.3}$$

Here the second term on the right hand side represents a tangential derivative, since

$$(\delta_\alpha^\beta - \varrho^{-2}\xi_\alpha\xi_\beta)\xi_\beta \equiv 0. \tag{7.4}$$

If we introduce the symbol D_α for $\partial/\partial x_\alpha$, we can write (7.3) in the form

$$D_\alpha = L'_\alpha + L''_\alpha \tag{7.5}$$

where

$$L'_\alpha = \varrho^{-2}\xi_\alpha\xi_\beta D_\beta, \quad L''_\alpha = (\delta^\beta_\alpha - \varrho^{-2}\xi_\alpha\xi_\beta) D_\beta. \tag{7.6}$$

Here for fixed ξ the expression L'_α and L''_α are linear homogeneous first order differential operators with constant coefficients. Moreover L''_α is of the form $c_\beta D_\beta$, where $c_\beta \xi_\beta = 0$. We call more generally any homogeneous differential operator, which has constant coefficients and is of some order m, *tangential* to the element (x, ξ), if L can be written in the form

$$L = c_{i\beta} D_\beta L_i, \tag{7.7}$$

where the L_i are homogeneous differential operators of order $m - 1$ with constant coefficients, and the $c_{i\beta}$ are constants satisfying

$$c_{i\beta}\xi_\beta = 0. \tag{7.8}$$

Here $\beta = 1, \ldots, n$, while i shall range over any finite set of positive integers. It is clear that, if L and M are any homogeneous operators with *constant* coefficients, and if L is tangential to (x, ξ), then also LM and ML are tangential (indeed $LM = ML = c_{i\beta} D_\beta L_i M$). Moreover, if L and M are of the same order, and both tangential, then also $L + M$ is tangential.

For a fixed ξ the operator L''_α is tangential to (x, ξ) in the sense defined here. It follows that

$$\begin{aligned} D_{\alpha_1} D_{\alpha_2} \ldots D_{\alpha_m} &= (L'_{\alpha_1} + L''_{\alpha_1})(L'_{\alpha_2} + L''_{\alpha_2}) \ldots (L'_{\alpha_m} + L''_{\alpha_m}) \\ &= L'_{\alpha_1} L'_{\alpha_2} \ldots L'_{\alpha_m} + M_{\alpha_1\alpha_2\ldots\alpha_m} \\ &= \varrho^{-2m}\xi_{\alpha_1}\xi_{\alpha_2}\ldots\xi_{\alpha_m}(\xi_\beta D_\beta)^m + M_{\alpha_1\alpha_2\ldots\alpha_m}, \end{aligned} \tag{7.9}$$

where $M_{\alpha_1\ldots\alpha_m}$ is a tangential operator. Since (7.9) is actually satisfied identically in ξ, we have that $M_{\alpha_1\ldots\alpha_m}$ is a homogeneous differential operator of degree m in D with coefficients, which are rational homogeneous functions of degree 0 in ξ. More precisely each coefficient multiplied by ϱ^{2m} is a form of degree $2m$ in ξ. As a tangential operator $M_{\alpha_1\ldots\alpha_m}$ can be put into the form

$$M_{\alpha_1\ldots\alpha_m} = c_{i\beta} D_\beta P_i, \tag{7.10}$$

where the P_i are differential operators of degree $m-1$, and the $c_{i\beta}$ satisfy (7.8). It is evident from (7.9) that each term of the expanded product contributing to $M_{\alpha_1\dots\alpha_m}$ contains at least one factor L''_α for some α. It follows that we actually have a representation of the form

$$M_{\alpha_1\dots\alpha_m} = (\delta^\beta_\alpha - \varrho^{-2}\xi_\alpha\xi_\beta)D_\beta P_{\alpha\alpha_1\alpha_2\dots\alpha_m}, \tag{7.12}$$

where $P_{\alpha\alpha_1\dots\alpha_m}$ is a homogeneous differential operator of order $m-1$ with coefficients which are rational in ξ, with denominator $\varrho^{2(m-1)}$ and a numerator, which is a form of degree $2(m-1)$.

We have for example for $m = 2$

$$\begin{aligned} D_{\alpha_1}D_{\alpha_2} = {} & \varrho^{-4}\xi_{\alpha_1}\xi_{\alpha_2}(\xi_\beta D_\beta)^2 \\ & + \tfrac{1}{2}(\delta^\alpha_\beta - \varrho^{-2}\xi_\alpha\xi_\beta)D_\beta[\delta^\alpha_{\alpha_1}(\delta^\gamma_{\alpha_2} + \varrho^{-2}\xi_{\alpha_2}\xi_\gamma) \\ & + \delta^\alpha_{\alpha_2}(\delta^\gamma_{\alpha_1} + \varrho^{-2}\xi_{\alpha_1}\xi_\gamma)]\,D_\gamma. \end{aligned} \tag{7.13}$$

Let now L be any linear differential operator of order m with variable coefficients. We can write L in the form

$$L = A_{\alpha_1\dots\alpha_m}D_{\alpha_1}D_{\alpha_2}\dots D_{\alpha_m} + L', \tag{7.14}$$

where L' is an operator of order $m-1$ at most. The *characteristic* form of L is defined by

$$Q = A_{\alpha_1\dots\alpha_m}\xi_{\alpha_1}\dots\xi_{\alpha_m} = Q(A, \xi). \tag{7.15}$$

Then by (7.9), (7.12)

$$\begin{aligned} L = {} & \varrho^{-2m}Q(A,\xi)(\xi_\beta D_\beta)^m + L' \\ & + A_{\alpha_1\dots\alpha_m}(\delta^\beta_\alpha - \varrho^{-2}\xi_\alpha\xi_\beta)D_\beta P_{\alpha\alpha_1\dots\alpha_m}. \end{aligned} \tag{7.16}$$

Here A stands for the set of coefficients $A_{\alpha_1\dots\alpha_m}$.

Consider now a quasi-linear system of equations

$$L_{ik}[v^k(x)] = B_i \qquad (i, k = 1, \dots, N) \tag{7.17}$$

for the dependent variables $v^1(x), \dots, v^N(x)$. Here L_{ik} shall be a linear homogeneous operator of order m_k, characteristic form Q_{ik}, and coefficients $A^{ik}_{\alpha_1\dots\alpha_{m_k}}$. Each L_{ik} can be represented in the form (7.16). Denote by $u_1(x), \dots, u_{N'}(x)$ the set formed by the $v^k(x)$ and their derivatives of orders $\leqq m_k - 1$, taken in some order. The A^{ik} and B_i are then known functions of the x_l and u_k. Equations (7.17) then take the form

$$(7.18)\quad \varrho^{-2m_k} Q_{\underline{i}k}(A^{\underline{i}k}, \xi)(\xi_\beta D_\beta)^{m_k}[v^k(x)] = B_i + c_{ir\,\beta}(A, \xi) D_\beta[u_r].$$

Here A denotes the set of all the coefficients $A^{ik}_{\alpha_1 \ldots \alpha_{m_k}}$ occurring in all L_{ik}, and A^{ik} denotes the sets of those occurring in a special L_{ik}. The $c_{ir\,\beta}(A, \xi)$ are linear in the A and rational in the ξ. More precisely they are *even* polynomials in the ξ_α/ϱ. Moreover the $c_{ir\,\beta}(A, \xi)$ satisfy the relations

$$(7.19)\qquad c_{ir\,\beta}(A, \xi)\xi_\beta = 0$$

identically in A and ξ. The *characteristic form* of the system (7.17) is given by

$$(7.20)\qquad Q(A, \xi) = \text{determinant } Q_{\underline{i}\underline{k}}(A^{\underline{i}\underline{k}}, \xi).$$

It is homogeneous of degree N in A and of degree

$$\sum_{k=1}^{N} m_k$$

in ξ. For values of A and ξ, for which $Q(A, \xi) \neq 0$ we can form the reciprocal matrix $Q^{ik}(A, \xi)$ to the matrix of the $Q_{\underline{i}\underline{k}}(A^{\underline{i}\underline{k}}, \xi)$. Each $Q^{ik}(A, \xi)$ is a quotient with denominator $Q(A, \xi)$ and a numerator, which is homogeneous of degree $N - 1$ in A and of degree

$$\sum_{s=1}^{N} m_s - m_i$$

in ξ. Then (7.18) yields

$$\varrho^{-2m_{\underline{s}}}(\xi_\beta D_\beta)^{m_{\underline{s}}}[v^{\underline{s}}(x)] = Q^{si}(A, \xi)(B_i + c_{ir\,\beta}(A, \xi) D_\beta[u_r]).$$

Any first derivative of any of the u_i is either another u_k or is a derivative of order m_s of some v^s. In the latter case we have, say,

$$\begin{aligned} u_{i,\beta} &= D_{\alpha_1} \ldots D_{\alpha_{m_{\underline{s}}}} v^{\underline{s}} \\ &= \varrho^{-2m_{\underline{s}}} \xi_{\alpha_1} \ldots \xi_{\alpha_{m_{\underline{s}}}} (\xi_\beta D_\beta)^{m_{\underline{s}}}[v^{\underline{s}}] + M_{\alpha_1 \ldots \alpha_{m_{\underline{s}}}}[v^{\underline{s}}] \\ &= \xi_{\alpha_1} \ldots \xi_{\alpha_{m_{\underline{s}}}} Q^{\underline{s}i}(A, \xi)(B_i + c_{ir\,\beta}(A, \xi) D_\beta[u_r]) \\ &\qquad\qquad + (\delta^\beta_\alpha - \varrho^{-2}\xi_\alpha \xi_\beta) D_\beta P_{\alpha\alpha_1 \ldots \alpha_{m_{\underline{s}}}}[v^{\underline{s}}] \\ &= S_{ir\,\beta\alpha} u_{r,\alpha} + T_{i\beta}. \end{aligned}$$

Here $T_{i\beta} = T_{i\beta}(A, B, \xi)$ is rational in its arguments. It is in fact linear in B with coefficients, which multiplied by $Q(A, \xi)$ are polynomials in A and ξ. The expressions $S_{ir\beta\alpha} = S_{ir\beta\alpha}(A, \xi)$ are also rational in their arguments. Multiplied with $Q(A, \xi)$ the $S_{ir\alpha\beta}$ reduce to polynomials in the A, ξ, $\xi_\alpha \xi_\beta / \varrho^2$. Moreover the $S_{ir\alpha\beta}$ and $T_{i\beta}$ are homogeneous of degree 0 in ξ and satisfy

$$S_{ir\beta\alpha}\xi_\alpha = 0$$

identically in ξ and A. In the case, where $u_{i,\beta}$ is another u_k, we have trivially for $u_{i,\beta}$ an expression $S_{ir\beta\alpha}u_{r,\alpha} + T_{i\beta}$ with $S_{ir\beta\alpha} = 0$ and $T_{i\beta} = u_k$.

In this way we have that for any solution system v^k of (7.17) the derivatives u_i satisfy a system of first order equations, which is canonical. The coefficients $T_{i\beta}$ and $S_{ir\beta\alpha}$ are *universal functions* of the original coefficients $A^{ik}_{\alpha_1 \ldots \alpha_{m_s}}$, B_i and of the u_r and ξ, which depend otherwise only on the integers m_k, N, n. If the system (7.17) was quasi-linear, i.e. if the A^{ik} and B_i were known functions of the x and v^k and of their derivatives of orders $\leqq m_k - 1$, the $T_{i\beta}$ and $S_{ir\beta\alpha}$ will be known functions of the x, u, ξ. If the original system (7.17) was linear, the $S_{ir\beta\alpha}$ will be known functions of x and ξ, and the $T_{i\beta}$ will be linear functions of the u_k with coefficients, which are known functions of the x and ξ.

The system (7.17) is elliptic, if $Q(A, \xi) \neq 0$ for all real $\xi \neq 0$ and all values of the A, which occur. (In the quasi-linear case elliptic character might depend on the special solutions v^k considered.) For an elliptic system (7.17) the corresponding canonical system for the u_i has coefficients, which have no real poles for $\xi \neq 0$.

A reduction of a *determined* system of differential equations (7.17) to an *over-determined* one (7.2) would be of little advantage in establishing *existence* of solutions with given boundary data, since obviously data must be much more restricted for the over-determined system. However we shall only use this reduction in establishing *regularity properties* of solutions v^k of (7.17), whose existence is assumed.

Reduction of an elliptic system of higher order equations to a *determined and elliptic* system of first order equations would be much more difficult than the reduction to an elliptic canonical system given here.

Such a reduction to canonical form can be carried out not only for quasi-linear systems of equations, but for general non-linear ones as well. We shall just consider here the case of a general m-th order equation for a single unknown function $u = u(x)$. We denote by $p_{i_1 \ldots i_k}$ the derivative $D_{i_1} \ldots D_{i_k} u$ of u. Let p denote the set of functions consisting of u and all its derivatives of order $\leqq m$. The differential equation can then be written in the form

$$F(x, p) = 0. \tag{7.21}$$

Differentiating with respect to x_i we obtain the relation

$$\begin{aligned} 0 &= F_{x_i} + \sum_{k=0}^{m-1} \sum_{\substack{i_1, \ldots, i_k \\ =1, \ldots, n}} F_{p_{i_1 \ldots i_k}} p_{i_1 \ldots i_k i} \\ &\qquad + \sum_{\substack{i_1 \ldots, i_m \\ =1, \ldots, n}} F_{p_{i_1 \ldots i_m}} p_{i_1 \ldots i_m i} \\ &= F'_i(x, p) + \sum_{i_1, \ldots, i_m} F_{p_{i_1 \ldots i_m}} D_{i_1} \ldots D_{i_m} D_i u. \end{aligned}$$

We multiply these equations with ξ_i, summing over i and substituting

$$D_{i_1} \ldots D_{i_m} u = \varrho^{-2m} \xi_{i_1} \ldots \xi_{i_m} (\xi_\beta D_\beta)^m [u] + M_{i_1 \ldots i_m}[u].$$

Putting

$$Q(x, p, \xi) = \sum_{i_1, \ldots, i_m} F_{p_{i_1 \ldots i_m}}(x, p) \, \xi_{i_1} \ldots \xi_{i_m}$$

we obtain an identity of the form

$$\begin{aligned} 0 &= \varrho^{-2m} Q(x, p, \xi)(\xi_\beta D_\beta)^{m+1}[u] + \xi_i F'_i(x, p) \\ &\qquad + F_{p_{i_1 \ldots i_m}}(x, p)(\delta^\beta_\alpha - \varrho^{-2} \xi_\alpha \xi_\beta) D_\beta (\xi_\gamma D_\gamma) P_{\alpha i_1 \ldots i_m}[u] \\ &= \varrho^{-2m} Q(x, p, \xi)(\xi_\beta D_\beta)^{m+1}[u] + \xi_i F'_i(x, p) + R_{i\beta}(x, p, \xi) D_\beta p^i, \end{aligned} \tag{7.22}$$

where $p^1, \ldots, p^N$ denote the functions p in some order. Here $(1/\varrho) R(x, p, \xi)$ is an odd polynomial in the ξ_α/ϱ with $R_{i\beta} \xi_\beta \equiv 0$.

For $Q(x, p, \xi) \neq 0$ we can solve for $(\xi_\beta D_\beta)^{m+1}[u]$ in (7.22).

Now a first derivative of any p^i is either a p^k or is a derivative of u of order $m + 1$. In the latter case it can be expressed by (7.9) in terms of $(\xi_\beta D_\beta)^{m+1}u$ and of tangential derivatives of u of order $m + 1$. Expressing the normal derivatives $(\xi_\beta D_\beta)^{m+1}[u]$ by (7.22) we again obtain a canonical system of differential equations for the p^i. If the original equation (7.21) was elliptic, i.e. if for the x and p in question

$$Q(x, p, \xi) \neq 0 \text{ for all real } \xi \neq 0,$$

then the resulting canonical system will be elliptic.

As a concrete example we consider the general second order differential equation in two independent variables

$$F(x, y, u, p, q, r, s, t) = 0, \tag{7.23}$$

where $p = u_x$, $q = u_y$, $r = u_{xx}$, $s = u_{xy}$, $t = u_{yy}$. We write ξ, η for ξ_1, ξ_2, and ϱ^2 for $\xi^2 + \eta^2$. The characteristic form is

$$Q = F_r \xi^2 + F_s \xi\eta + F_t \eta^2. \tag{7.24}$$

To reduce the resulting canonical system for u, p, q, r, s, t to manageable proportions we introduce a number of abbreviations. We use the notation $\overline{D}_x$, $\overline{D}_y$ for the tangential components of the operators D_x, D_y:

$$\overline{D}_x = D_x - \varrho^{-2}\xi(\xi D_x + \eta D_y), \quad \overline{D}_y = D_y - \varrho^{-2}\eta(\xi D_x + \eta D_y).$$

Moreover we put

$$\begin{aligned}
F' &= \xi(F_x + pF_u + rF_p + sF_q) + \eta(F_y + qF_u + sF_p + tF_q) \\
A &= \xi(1 + \varrho^{-2}\xi^2)r + \eta(1 + 2\varrho^{-2}\xi^2)s + \varrho^{-2}\xi\eta^2 t \\
&= (D_x(\xi D_x + \eta D_y) + \varrho^{-2}\xi(\xi D_x + \eta D_y)^2)\,[u] \\
B &= \varrho^{-2}\xi^2\eta r + \xi(1 + 2\varrho^{-2}\eta^2)s + \eta(1 + \varrho^{-2}\eta^2)t \\
&= (D_y(\xi D_x + \eta D_y) + \varrho^{-2}\eta(\xi D_x + \eta D_y)^2)\,[u].
\end{aligned}$$

Then the canonical system of equations becomes

$$u_x = p, \; u_y = q, \; p_x = r, \; p_y = q_x = s, \; q_y = t \tag{7.25}$$

$$r_x = \bar{D}_x r - \frac{F'}{Q}\varrho^{-2}\xi^3 - \frac{\varrho^{-2}\xi^3}{Q}\{(F_r - \xi^{-2}Q)\bar{D}_x A + \tfrac{1}{2}F_s(\bar{D}_x B + \bar{D}_y A) + F_t \bar{D}_y B\}$$

$$r_y = s_x = \bar{D}_x s - \frac{F'}{Q}\varrho^{-2}\xi^2\eta$$

$$-\frac{\varrho^{-2}\xi^2\eta}{Q}\{F_r\bar{D}_x A + \tfrac{1}{2}(F_s - \xi^{-1}\eta^{-1}Q)(\bar{D}_x B + \bar{D}_y A) + F_t\bar{D}_y B\}$$

with two additional equations for $s_y = t_x$ and t_y obtained by permutation from the last ones.

The formula for integration by parts on a sphere

Let $u(x)$ and $a_i(x)$ for $i = 1, \ldots, n$ be functions of class C_1 in the sphere $|x - z| \leqq r$. Then

$$\int_{|x-z|<r} a_i(x)u_{,i}(x)\,dx = -\int_{|x-z|<r} a_{i,i}(x)u(x)\,dx + \frac{1}{r}\int_{|x-z|=r} a_i(x)(x_i - z_i)u(x)\,dS_x.$$

Differentiation with respect to r yields

$$\text{(7.26)} \quad \int_{|x-z|=r} a_i(x)u_{,i}(x)\,dS_x = -\int_{|x-z|=r} a_{i,i}(x)u(x)\,dS_x + \frac{d}{dr}\frac{1}{r}\int_{|x-z|=r} a_i(x)(x_i - z_i)u(x)\,dS_x.$$

Since (7.26) only involves the values of the a_i and u and of their first derivatives on the sphere $|x - z| = r$, that formula stays valid, if the a_i and u are only defined and of class C_1 in a neighborhood of the surface $|x - z| = r$. They can always be continued into the full interior as functions of class C_1. In particular we obtain the relation

$$\text{(7.27)} \quad \int_{|x-z|=r} a_i(x)u_{,i}(x)\,dS_x = -\int_{|x-z|=r} a_{i,i}(x)u(x)\,dS_x$$

whenever the a_i and u are of class C_1 in a neighborhood of $|x - z| = r$, provided the $a_i(x)$ satisfy the identity

$$a_i(x)(x_i - z_i) = 0 \tag{7.28}$$

in that neighborhood. Condition (7.28) means that $a_i u_{,i}$ shall be a tangential derivative of u on the spheres about z.

Spherical integrals of solutions of a canonical system

Let the functions $u_i(x)$ be solutions of class C_1 of a *linear* elliptic canonical system (7.2). Let $b(x, \xi)$ and $a_i(x, \xi)$ be functions of class C_1 in x and ξ for $\xi \neq 0$. We form the expression

$$J(z, r) = \int_{|x-z|=r} [a_i(x, x - z)\, u_i(x) + b(x, x - z)]\, dS_x. \tag{7.29}$$

We shall show that the first derivatives of J are expressions of the same form.

We can write J in the form

$$J(z, r) = r^{n-1} \int_{\Omega_\eta} [a_m(z + r\eta, r\eta) u_m(z + r\eta) + b(z + r\eta, r\eta)]\, d\omega_\eta \tag{7.30}$$

from which it follows that

$$\frac{\partial J(z, r)}{\partial z_\alpha} = \int_{|x-z|=r} [a_{m,\alpha}(x, x-z) u_m(z) + a_m(x, x-z) u_{m,\alpha}(x) + b_{,\alpha}(x, x-z)]\, dS$$

$$\frac{\partial J(z, r)}{\partial r} = \frac{n-1}{r} J(z, r)$$

$$+ \frac{1}{r} \int_{|x-z|=r} (a_{m,\alpha} u_m + a_{m;\alpha} u_m + a_m u_{m,\alpha} + b_{,\alpha} + b_{;\alpha})(x_\alpha - z_\alpha)\, dS_x.$$

Here $a_{m,\alpha}$ and $a_{m;\alpha}$ denote respectively the derivatives of $a_m(x, \xi)$ with respect to x_α and ξ_α taken for $\xi = x - z$. We have assumed that the canonical system is *linear*. Then its coefficients are of the form

$$S_{mi\alpha\beta} = S_{mi\alpha\beta}(x, \xi), \quad T_{m\alpha} = T'_{mk\alpha}(x, \xi)\, u_k(x) + T''_{m\alpha}(x, \xi). \tag{7.30a}$$

Then the canonical equations yield for $\xi = x - z$

$$u_{m,\alpha}(x) = S_{ms\alpha\beta}(x, x-z)u_{s,\beta}(x) + T'_{mi\alpha}(x, x-z)u_i(x) + T''_{m\alpha}(x, x-z)$$

where

$$S_{ms\alpha\beta}(x, x-z)(x_\beta - z_\beta) = 0 \text{ for } x \neq z.$$

Substituting this expression for $u_{m,\alpha}(x)$, we can apply formula (7.27) for integration by parts, and obtain for the first derivatives of J expressions of the form

$$\text{(7.31a)} \quad \frac{\partial J(z, r)}{\partial z_\alpha} = \int\limits_{|x-z|=r} [a'_i(x, x-z)u_i(x) + b'(x, x-z)]dS_x$$

$$\text{(7.31b)} \quad \frac{\partial J(z, r)}{\partial r} = \int\limits_{|x-z|=r} [a''_i(x, x-z)u_i(x) + b''(x, x-z)]dS_x,$$

where

$$\begin{aligned} \text{(7.32a)} \quad a'_i(x, \xi) = {} & a_{i,\alpha}(x, \xi) - a_m(x, \xi)S_{mi\alpha\beta,\beta}(x, \xi) \\ & - a_m(x, \xi)S_{mi\alpha\beta;\beta}(x, \xi) - a_{m,\alpha}(x, \xi)S_{mi\alpha\beta}(x, \xi) \\ & - a_{m;\alpha}(x, \xi)S_{mi\alpha\beta}(x, \xi) + a_m(x, \xi)T'_{mi\alpha}(x, \xi) \end{aligned}$$

$$\text{(7.32b)} \quad b'(x, \xi) = a_m(x, \xi)\, T''_{m\alpha}(x, \xi) + b_{,\alpha}(x, \xi)$$

$$\begin{aligned} \text{(7.32c)} \quad a''_i(x, \xi) = \frac{1}{|\xi|} [& a_{i,\alpha}\xi_\alpha + a_{i;\alpha}\xi_\alpha + (n-1)a_i \\ & - a_{m,\beta}S_{mi\alpha\beta}\xi_\alpha - a_{m;\beta}S_{mi\alpha\beta}\xi_\alpha - a_m S_{mi\alpha\beta,\beta}\xi_\alpha \\ & - a_m S_{mi\alpha\beta;\beta}\xi_\alpha - a_m S_{mi\alpha\alpha} + a_m T'_{mi\alpha}\xi_\alpha] \end{aligned}$$

$$\text{(7.32d)} \quad b''(x, \xi) = \frac{1}{|\xi|} [b_{,\alpha}\xi_\alpha + b_{;\alpha}\xi_\alpha + (n-1)b + a_m T''_{m\alpha}\xi_\alpha].$$

Differentiability of solutions of linear elliptic systems

Formulae (7.31a, b) give an expression for the first derivatives of an integral J of the form (7.29) as an expression of the same form. Here the u_i are assumed to be solutions of the canonical system (7.2), which must be elliptic, since otherwise the $T_{m\alpha}$

and $S_{mr\alpha\beta}$ would not be defined for all directions ξ of the normal of a sphere. Formulae (7.31a, b) have been derived under the assumption that the u_i are in C_1. It has also been assumed that a_i, b, $S_{mi\alpha\beta}$ are in C_1 and that $T''_{m\alpha}$ and $T'_{mi\alpha}$ are continuous.

By iteration of the procedure we get expressions for the higher derivatives of $J(z, r)$ with respect to the z_i and r, which are of the same form as J itself, but involve higher derivatives of the a_i, b, $S_{mi\alpha\beta}$, $T'_{mi\alpha}$, $T''_{m\alpha}$. If the $a_i(x, \xi)$, $b(x, \xi)$, $S_{mi\alpha\beta}(x, \xi)$ are of class C_t in x and ξ for $\xi \neq 0$, and $T''_{m\alpha}(x, \xi)$, $T'_{mi\alpha}(x, \xi)$ of class C_{t-1}, it follows that $J(z, r)$ is of class C_t in z and r for $r > 0$. We can take for J in particular the spherical average $I_i(z, r)$ of $u_i(x)$, which corresponds to the choice

$$a_k(x, \xi) = \frac{1}{\omega_n} |\xi|^{1-n} \delta_i^k, \qquad b(x, \xi) = 0. \tag{7.33}$$

Since the a_k and b are analytic for $\xi \neq 0$, we have the theorem that the spherical averages of a solution u_i of class C_1 of a linear elliptic canonical system

$$u_{m,\alpha}(x) = S_{mi\alpha\beta}(x, \xi) u_{i,\beta}(x) + T'_{mi\alpha}(x, \xi) u_i(x) + T''_{m\alpha}(x, \xi) \tag{7.34}$$
$$(S_{mi\alpha\beta}(x, \xi)\xi_\beta \equiv 0)$$

are of class C_t for $r > 0$, if the $S_{mi\alpha\beta}(x, \xi)$ are of class C_t in x and the $T'_{mi\alpha}(x, \xi)$, $T''_{m\alpha}(x, \xi)$ of class C_{t-1} in x. (It is always assumed that the coefficients are rational homogeneous of degree 0 in ξ.)

We make use now of the theorem proved in Chapter IV pp. 86, 88: If the spherical means $I(z, r)$ of a continuous function $u(x) = u(x_1, \ldots, x_n)$ are of class C_t for $r > 0$, the function $u(x)$ is of class $C_{t-2[n/2]}$. (Here $[\alpha]$ is defined as the largest integer not exceeding α.) We thus have the theorem:

If the $u_i(x)$ form a solution of class C_1 of a linear elliptic canonical system (7.34) with coefficients $S_{mi\alpha\beta}(x, \xi)$ of class C_t in x and coefficients $T'_{mi\alpha}(x, \xi)$, $T''_{m\alpha}(x, \xi)$ of class C_{t-1} in x, then the $u_i(x)$ are of class $C_{t-2[n/2]}$.

We saw that a linear elliptic system of differential equations (7.17) for functions $v^1(x), \ldots, v^N(x)$ can be reduced to a linear elliptic system of canonical equations for the derivatives of the $v^i(x)$ of orders $\leqq m_i - 1$. The coefficients of the canonical system are rational in ξ, in the coefficients of the L_{ik} and in the B_i. We then have the following consequence of the last theorem:

Let

$$L_{ik}[v^k(x)] = B_i(x) \tag{7.35}$$

be a linear elliptic system of N equations for N functions $v^k(x)$. Let the order of the operator L_{ik} be m_k. Let the B_i and the coefficients of all the L_{ik} be of class C_t in x. Let each $v^k(x)$ be of class C_{m_k}. Then v^k is also of class C_{μ_k}, where

$$\mu_k = m_k - 1 + t - 2[n/2].\text{[60]} \tag{7.36}$$

Differentiability of solutions of non-linear elliptic systems

We consider the general quasi-linear canonical system (7.2) for functions $u_i(x)$. The coefficients $T_{m\alpha}$ and $S_{mr\alpha\beta}$ are then prescribed functions of the u_k and of x and ξ. They are rational in ξ, satisfying (7.1) for all values of x and of the u_i of a domain in xu-space, and for all $\xi \neq 0$. For the solution u_i considered the system shall be elliptic, i.e. the $T_{m\alpha}$ and $S_{mr\alpha\beta}$ as functions of ξ shall have no real pole $\xi \neq 0$, if we substitute for the u_i the values $u_i(x)$ corresponding to the solution under consideration.

By differentiation of (7.2) we can get canonical systems for the higher derivatives of the u_i, provided the u_i, $T_{m\alpha}$, $S_{mr\alpha\beta}$ are sufficiently often differentiable. We assume that the $T_{m\alpha}$, $S_{mr\alpha\beta}$ are of class C_p in their arguments x, u_i, and that the u_i are of class C_{p+1}. Differentiating (7.2) p-times with respect to the x_i we obtain an equation of the form

$$u_{m,\alpha_1\ldots\alpha_p\alpha} = \overline{T}_{m\alpha_1\ldots\alpha_p\alpha} + S_{mr\alpha\beta}u_{r,\alpha_1\ldots\alpha_p\beta} \tag{7.37}$$

where $\overline{T}_{m\alpha_1\ldots\alpha_p\alpha}$ is formed from x-derivatives of the $T_{m\alpha}(x, u(x), \xi)$ and $S_{mr\alpha\beta}(x, u(x), \xi)$. The expression $\overline{T}_{m\alpha_1\ldots\alpha_p\alpha}$ is a polynomial in the derivatives of the u_i of orders $\leqq p$ with coefficients which

[60] For more precise theorems for the case of systems of second order equations see Morrey [3].

are known functions of the x, u_k and rational homogeneous functions of degree 0 in the ξ. More precisely $\overline{T}_{m\alpha_1\ldots\alpha_p\alpha}$ is linear in the derivatives of the u_k of orders $> (p+1)/2$ with coefficients, which contain only derivatives of the u_k of orders $\leqq (p+1)/2$. We introduce all x-derivatives of the u_i of orders q with $(p+1)/2 < q \leqq p$ as new dependent variables v^k. Then we have in (7.37) a *linear* canonical system for the v^k (if we add the trivial equations expressing a derivative of a u_i of order $< p$ as a derivative of u_k of order $\leqq p$). The coefficients in this linear canonical system for the v^k are polynomials in the derivatives of the u_i of orders $\leqq (p+1)/2$ with coefficients depending on the x, u_k, ξ. Assume that the $T_{m\alpha}$ and $S_{mr\alpha\beta}$ are of class C_σ in x and the u_i, where

$$\sigma \geqq \left[\frac{3p+2}{2}\right].$$

Then the coefficients of the v^k in the linear canonical system (7.37) are of class $C_{[(p+2)/2]}$ and the coefficients $S_{mr\alpha\beta}$ of the first derivatives of the v^k are of class C_{p+1} since the u_i are of class C_{p+1}. If $p \geqq 2$, it follows from the theorem on linear canonical systems (p. 138) with $t = [(p+4)/2]$ that the v^k are of class $C_{[(p+4)/2]-2[n/2]}$. Since the v^k include the derivatives of the u_i of order p, it follows that solutions u_i of (7.2) of class C_{p+1} are of order C_μ with

$$\mu = \left[\frac{3p+4}{2}\right] - 2\left[\frac{n}{2}\right] \tag{7.38}$$

provided p is an integer with

$$p \geqq 2, \quad \left[\frac{3p+2}{2}\right] \leqq \sigma. \tag{7.39}$$

We have from (7.38) $\mu \geqq p + 2$ for

$$p \geqq 4\left[\frac{n}{2}\right]. \tag{7.40}$$

Thus under the conditions (7.39), (7.40) we have that for a solution system u_i of class C_{p+1} the u_i are also of class C_{p+2}. We arrive in this way at the following theorem:

Let (7.2) *be a quasi-linear canonical system of differential equations with coefficients* $T_{m\alpha}$, $S_{mr\alpha\beta}$ *which are of class* C_σ *in their arguments* x, u_i. *Let the* $u_i(x)$ *form a system of solutions of* (7.2) *of class* $C_{4[n/2]+1}$, *the canonical equations being elliptic for the* u_i. *Then the* u_i *are of class* C_μ, *for any* μ *for which*

$$\mu < \frac{2}{3}\sigma + 2. \tag{7.41}$$

A quasi-linear system of equations of the form (7.17) can again be reduced to a quasi-linear canonical system for the derivatives of the v^k of order $\leqq m_k - 1$. The coefficients of the canonical system are rational functions of the original coefficients and of the ξ. In this way we obtain the theorem:

Let

$$L_{ik}[v^k(x)] = B_i \tag{7.42}$$

be a quasi-linear system of N *equations for* N *functions* $v^k(x)$. *Let the operator* L_{ik} *be linear in the derivatives of order* m_k. *The coefficients of all the* L_{ik} *and the* B_i *shall depend on* x *and the derivatives of each* v^r *of order* $\leqq m_r - 1$. *Let the coefficients of the* L_{ik} *and the* B_i *be of class* C_σ *in their arguments. Let the* $v^k(x)$ *form a solution of* (7.42), *for which that system is elliptic, each* $v^r(x)$ *being of class* $C_{m_r+4[n/2]}$. *Then* v_r *is of class* $C_{m_r-1+\mu}$, *where* μ *is any number satisfying* (7.41).

We saw that a single general non-linear elliptic differential equation of order m for a function $u(x)$ can be reduced to a quasi-linear elliptic canonical system for the derivatives of u of orders $\leqq m$. The system is obtained by forming first derivatives of the original equations. In this way we obtain the following result: [61]

Let there be given a general non-linear differential equation

$$F(x, p) = 0, \tag{7.43}$$

where p *stands for the set of functions consisting of a function* $u(x)$ *and its* x*-derivatives of orders* $\leqq m$. *Let* F *be a function of class*

[61] See E. Hopf [1], C. B. Morrey [1], L. Nirenberg [1], [2] for stronger theorems in the case $m = 2$.

C_σ in its arguments x, p. Let $u(x)$ be a solution of (7.43) of class $C_{m+4[n/2]+1}$, for which (7.43) is an elliptic equation. Then $u(x)$ is also of class $C_{m+\mu}$ for any μ with

$$\mu < \frac{2}{3}(\sigma + 4). \tag{7.44}$$

Analyticity of solutions of linear elliptic systems analytic coefficients

Let the $u_i(x)$ be solutions of the *linear* elliptic canonical system

$$u_{m,\alpha}(x) = S_{ms\alpha\beta}(x, \xi)u_{s,\beta}(x) + T'_{mi\alpha}(x, \xi)u_i(x) + T''_{m\alpha}(x, \xi) \tag{7.45}$$
$$(i, m, s = 1, \ldots, N; \ \alpha, \beta = 1, \ldots, n).$$

We know that for coefficients $S_{ms\alpha\beta}$, $T'_{mi\alpha}$, $T''_{m\alpha}$ in C_∞ any solution u_i of (7.45) of class C_1 will also be in C_∞. We shall prove that if the coefficients are analytic in x, so is the solution.

For this purpose we consider an integral $J(z, r)$ of the form (7.29) formed from the u_i, with the help of functions $a_i(x, \xi)$ and $b(x, \xi)$. By (7.31a) $\partial J/\partial z_\alpha$ is an expression of the same form, only formed with coefficients $a'_i(x, \xi)$ and $b'(x, \xi)$. Here by (7.32a, b) the a'_i and b' are formed by applying certain linear first order differential operators to the a_i and b, the coefficients of those operators being combinations of the $S_{mi\alpha\beta}$, $T'_{mi\alpha}$, $T''_{m\alpha}$ and their first derivatives. Put $b(x, \xi) = a_{N+1}(x, \xi)$. Then the formulae connecting the a'_i, b' with the a_i, b are of the form

$$a'_i(x, \xi) = L_{ik\alpha}[a_k(x, \xi)] \quad (i, k = 1, \ldots, N+1; \alpha = 1, \ldots, n). \tag{7.46}$$

Here the $L_{ik\alpha}$ are first order differential operators containing differentiations with respect to the x and ξ variables, with coefficients, which are analytic functions of x, ξ for x restricted to a domain D and $\xi \neq 0$. [The suffix α refers to the z_α-derivative on the left hand side of (7.31a).]

To get estimates for the derivatives of J we compare the recursion formula (7.46) with the system of partial differential equations

$$\frac{\partial A_i(x, \xi, \eta, t)}{\partial t} = \eta_\alpha L_{ik\alpha}[A_k(x, \xi, \eta, t)] \tag{7.47}$$

with initial conditions

$$A_i = \frac{1}{\omega_n} | \xi |^{1-n} \delta^i_s \text{ for } t = 0, \tag{7.48}$$

where s is one of the numbers $1, \ldots, N$ and is fixed. For any real x, ξ, η, t with x in D and $\xi \neq 0$ the coefficients of the differential equations (7.47) for the A_i and the initial data (7.48) are analytic functions. Let D' be any compact subset of the domain D. There exists another compact subset D'' of D, such that D' is interior to D''. Let r be a positive number such that the r-neighborhood of any point of D' is in D''. By the theorem of Cauchy and Kowalewski we can find a solution system $A_i(x, \xi, \eta, t)$ of (7.47), (7.48), which is regular analytic in x, ξ, η, t in real ε-neighborhood of the set

$$| x | \subset D'', \ | \xi | = r, \ | \eta | = 1, \ t = 0$$

in $x\xi\eta t$-space.

Let $u_i(x)$ be the solution under consideration of the canonical system (7.45). The $u_i(x)$ shall be of class C_1 in D. We form the expression

$$J(z, \eta, t) = \int\limits_{|x-z|=r} \left[\sum_{i=1}^{N} A_i(x, x-z, \eta, t) u_i(x) + A_{N+1}(x, x-z, \eta, t) \right] dS_x.$$

The expression $J(z, \eta, t)$ is defined for z in D', $| \eta | = 1$, $| t | < \varepsilon$, and is analytic in η and t. By (7.31a), (7.46), (7.47) we have

$$\frac{\partial J(z, \eta, t)}{\partial t} = \eta_\alpha \frac{\partial J(z, \eta, t)}{\partial z_\alpha}$$

$$J(z, \eta, 0) = \frac{1}{\omega_n} r^{1-n} \int\limits_{|x-z|=r} u_s(x) \, dS_x = I_s(z, r).$$

Hence

$$J(z, \eta, t) = J(z + t\eta, \eta, 0) = I_s(z + t\eta, r). \tag{7.49}$$

Since $J(z, \eta, t)$ is analytic in η, t for $|\eta| = 1$, $|t| < \varepsilon$, it follows from (7.49) that $I_s(z, r)$ is analytic in z for z in D'. We see in this way that the spherical average $I_s(z, r)$ of $u_s(x)$ is an analytic function of z for any sphere with center z and radius r contained in D. Similarly the first n r-derivatives of $I_s(z, r)$ are analytic functions of z uniformly in r, for r sufficiently small and bounded away from 0. It follows then from formulae (4.24), (4.26) that $u_s(x)$ itself is an analytic function of x in D.

Consequently the solutions $u_i(x)$ of a linear elliptic canonical system with analytic coefficients are themselves analytic. Using the reduction of linear elliptic systems of the form (7.35) to canonical systems, we have the theorem:

Let (7.35) *be a linear elliptic system of N equations for N unknown functions $v^k(x)$. Let the $B_i(x)$ and the coefficients of the L_{ik} be analytic in a domain D. Let L_{ik} have the order m_k. Then any solution system $v^i(x)$, for which each $v^k(x)$ is of class C_{m_k}, consists of functions, which are analytic in D.*

For quasi-linear elliptic systems with coefficients, which are of class C_∞ in their arguments, it has been established in the preceding section that all solutions of class $C_{4[n/2]+1}$ are of class C_∞. It would be tempting to give a proof of the analyticity of the solutions in the case of analytic coefficients by establishing sufficiently strong estimates for the derivatives, whose existence has been proved already. It appears doubtful however that *sufficiently good* estimates can be obtained in the non-linear case (in the manner used here for linear systems). One reason is that in the non-linear case the expressions (4.24) or (4.26) for the u_i in terms of their iterated spherical means have to be used infinitely often (not only once, as in the linear case) in order to express derivatives of the u_i of arbitrarily high order in terms of a fixed number of derivatives.[62]

[62] For the general proof of the analyticity of solutions of non-linear analytic equations see Petrovskii [2], [3].

Differentiability of continuous weak solutions of a linear elliptic equation [63]

Let L be a linear elliptic differential operator of order m. Then L can be written in the form (3.1) with coefficients $A_{i_1 \ldots i_k}(x)$. The characteristic form $Q(x, \xi)$ of L satisfies (3.7). We assume that there exists a non-negative integer s such that the coefficients $A_{i_1 \ldots i_k}(x)$ of the derivatives of order k are of class C_{s+k} in the domain D. The adjoint differential operator $\bar{L}$ is defined by (3.37). Since the order m of the elliptic operator L is necessarily an even number, we have from (3.37) that L and $\bar{L}$ have the same characteristic form $Q(x, \xi)$. It is also evident from (3.37) that the coefficients of the k-th order derivatives in $\bar{L}$ are again of class C_{s+k}.

Let $u(x)$ be a solution of class C_m of the equation

$$L[u(x)] = f(x). \tag{7.50}$$

Let R be a closed region in D with boundary S. We have from (3.36)

$$\int_R (u(x)\bar{L}[v(x)] - f(x)v(x))\,dx = 0 \tag{7.51}$$

for every $v(x)$ of class C_m, which vanishes with its derivatives of order $\leqq m - 1$ on S. Conversely, if u is in C_m and if (7.51) holds for all R in D and all v in C_m vanishing of order m on the boundary of R, then u will be a solution of (7.50) in D. It is even sufficient to restrict R to be a sphere in D and v to be a function of class C_∞ which vanishes outside a smaller concentric sphere.

This leads to a definition of "weak" solutions u of (7.50).

Definition: $u(x)$ is called a continuous *weak* solution of the differential equation (7.50) (in which $f(x)$ is assumed to be continuous), if $u(x)$ is continuous in D, and if for every solid sphere $|x - z| \leqq r$ in D

[63] See Friedrichs [2]; Weyl [1], p. 415; Višik [1]; Garding [2]; Schwartz [2], v. 1; Browder [4]; Browder [3], p. 232 (apparently the first complete proof of "Weyl's lemma" for general elliptic equations); Friedrichs[1]; John [9]. The theorems given by Browder, Friedrichs, John, though roughly of the same degree of generality, differ so much in the types of regularity assumptions made, that a precise comparison is difficult. They have been extended recently by P. D. Lax [1].

$$\int_{|x-z|\leq r} (u(x)\bar{L}[v(x)] - f(x)v(x))\,dx = 0 \tag{7.52}$$

for all $v(x)$ of class C_∞ vanishing outside a sphere of radius $< r$ about z.

A solution u of (7.50) of class C_m in the ordinary sense will be called a *strict* solution of (7.50). A continuous weak solution, which is of class C_m, is necessarily a strict solution. We are interested here in the converse problem: When are weak solutions strict solutions, or more generally, when do they possess derivatives? The line of approach followed in the last sections to establish existence of higher order derivatives of u by reduction to a canonical system is not so profitable here, since we would have to introduce the derivatives of u of orders $\leq m-1$ as new dependent variables. The existence of these derivatives is just one of the facts to be proved. However a modification of the arguments used previously can be made to yield results of the type desired. One only has to carry out more carefully the differentiation of spherical integrals, so as to make the existence of the derivatives evident.

Assume that $u(x)$ is a continuous weak solution of (7.50) in D. Let $v(x)$ be a function of class C_∞, which vanishes outside a sphere of radius R about the point z. Then (7.52) holds for all $r > R$ and hence also for $r = R$. Let next $w(x)$ be any function of class C_∞ in D. Let $\theta(s)$ denote some specific function of s of class C_∞ for all s, which satisfies the conditions

$$\theta(s) = 1 \text{ for } s < 0;\ \theta(s) = 0 \text{ for } s > 1. \tag{7.52a}$$

Then for any ε with $0 < \varepsilon < 1$ we have in

$$v_\varepsilon(x) = (r^2 - |x-z|^2)^m\,\theta(1 - \varepsilon^{-1} + r^{-2}\varepsilon^{-1}|x-z|^2)\,w(x)$$

a function of class C_∞ with

$$v_\varepsilon(x) = (r^2 - |x-z|^2)^m w(x) \quad \text{for } |x-z| < r\sqrt{1-\varepsilon}$$
$$v_\varepsilon(x) = 0 \quad \text{for } |x-z| > r.$$

Equation (7.52) will hold for $v = v_\varepsilon(x)$. Since $v_\varepsilon(x)$ and its derivatives of order $\leqq m$ are bounded uniformly in ε in the full sphere $|x - z| \leqq r$ and converge uniformly for $\varepsilon \to 0$ in every smaller sphere, it follows that

$$\int_{|x-z|<r} u(x)\overline{L}[(r^2 - |x-z|^2)^m w(x)]\,dx = \int_{|x-z|<r} (r^2 - |x-z|^2)^m w(x) f(x)\,dx \tag{7.53}$$

for every $w(x)$ of class C_∞. It holds then also for any $w(x)$ of class C_m, since such w can be approximated uniformly with their first m derivatives by functions in C_∞.

Let z be a fixed point of D. Denote by ϱ the distance $|x - z|$. Let $v(x)$ be a function of class C_{m+j} for some non-negative j, and let v vanish in a neighborhood of z, say

$$v(x) = 0 \text{ for } |x - z| < \varepsilon. \tag{7.54}$$

Then for any integer $\alpha \geqq m$

$$(r - \varrho)^\alpha v(x) = (r^2 - \varrho^2)^m w(x),$$

where

$$w(x) = (r - \varrho)^{\alpha - m}(r + \varrho)^{-m} v(x)$$

is of class C_{m+j}, even at $x = z$. Hence from (7.53)

$$\int_{\varrho<r} u(x)\overline{L}[(r-\varrho)^\alpha v(x)]dx = \int_{\varrho<r} (r-\varrho)^\alpha v(x) f(x)\,dx. \tag{7.55}$$

Given the linear m-th order differential operator $\overline{L}$ acting on functions $v(x)$ we can define a sequence of *derived operators* $\overline{L}^{(k)}$ (with respect to the fixed function $\varrho(x)$).[64] We first define the first derived operator $\overline{L}'$ as a commutator by the formula

$$\overline{L}'[v(x)] = \varrho\overline{L}[v] - \overline{L}[\varrho v]$$

and then generally $\overline{L}^{(k)}$ by

$$\overline{L}^{(0)} = \overline{L},\ \overline{L}^{(k+1)} = \varrho\overline{L}^{(k)} - \overline{L}^{(k)}\varrho = (L^{(k)})'. \tag{7.55a}$$

[64] See John [3], p. 99.

Since the highest order coefficients in the differential operators $\varrho\bar{L}$ and $\bar{L}\varrho$ are the same, the operator $\bar{L}'$ is of order $m-1$ at most. Similarly $L^{(k)}$ is at most of order $m-k$, and vanishes identically for $k > m$. The coefficients of the derivatives of order $< k$ in the expression for $\bar{L}$ make no contribution to $\bar{L}^{(k)}$. According to (3.37) the coefficients of $\bar{L}^{(k)}$ are linear combinations of the $A_{i_1\ldots i_\beta}(x)$ with $\beta \geqq k$ and of their derivatives of order $\leqq \beta - k$. Since $A_{i_1\ldots i_\beta}(x)$ had been assumed to be of class $C_{s+\beta}$ and $\varrho(x)$ is of class C_∞ for $\varrho \neq 0$, it follows that the coefficients of the operator $\bar{L}^{(k)}$ are of class C_{s+k} for $\varrho \neq 0$.

We have from the definition of $\bar{L}'$ for constant r

$$\bar{L}[(r-\varrho)v] = (r-\varrho)\bar{L}[v] + \bar{L}'[v].$$

From this follows by induction for non-negative integers α the more general identity

$$\bar{L}[(r-\varrho)^\alpha v] = \sum_{k=0}^{\alpha} \binom{\alpha}{k} (r-\varrho)^{\alpha-k}\bar{L}^{(k)}[v]. \tag{7.56}$$

If we take here for v the function $v = 1$, for α the value m, and evaluate the formula resulting from (7.56) on the surface $\varrho = r$ we find

$$\begin{aligned} \bar{L}^{(m)}[1] &= (\bar{L}[(r-\varrho)^m])_{\varrho=r} \\ &= m!\left(\sum_{i_1,\ldots,i_m} A_{i_1\ldots i_m}(x)\frac{\partial\varrho}{\partial x_{i_1}}\cdots\frac{\partial\varrho}{\partial x_{i_m}}\right)_{\varrho=r} \\ &= m!\,Q(x, x-z)\varrho^{-m}, \end{aligned}$$

where $Q(x, \xi)$ is the characteristic form of L. Since $\bar{L}^{(m)}$ is an operator of order 0 we have more generally

$$\bar{L}^{(m)}[v] = v\bar{L}^{(m)}[1] = m!\,Q(x, x-z)\varrho^{-m}v. \tag{7.57}$$

We substitute the expansion (7.56) for $\bar{L}[(r-\varrho)^\alpha v]$ into formula (7.55). Observing that $\bar{L}^{(k)}[v] = 0$ for $k > m$ and making use of the expression (7.57) for $\bar{L}^{(m)}[v]$ we find for $\alpha \geqq m$

$$(7.58)\quad \int_{\varrho<r} \binom{\alpha}{m}(r-\varrho)^{\alpha-m}\overline{L}^{(m)}[1]v(x)u(x)\,dx$$
$$=\sum_{k=0}^{m-1}\int_{\varrho<r} -\binom{\alpha}{k}(r-\varrho)^{\alpha-k}\overline{L}^{(k)}[v]u(x)\,dx$$
$$+\int_{\varrho<r}(r-\varrho)^{\alpha}v(x)f(x)\,dx.$$

Let $w(x)$ be any function of class C_m, which vanishes for $\varrho<\varepsilon$. *Since L is elliptic* we have in

$$v(x)=\frac{1}{\overline{L}^{(m)}[1]}w(x)=\varrho^m w(x)/m!Q(x,x-z)$$

again a function of class C_m vanishing for $\varrho<\varepsilon$. We use this function v in (7.58). Replacing α by $m+\alpha$ and k by $m-k$ we obtain the identity

$$(7.59)\quad \int_{\varrho<r} u\frac{(r-\varrho)^{\alpha}}{\alpha!}w\,dx=\sum_{k=1}^{m}\int_{\varrho<r} u\frac{(r-\varrho)^{\alpha+k}}{(\alpha+k)!}\overline{L}_k[w]\,dx$$
$$+\int_{\varrho<r} f\frac{m!(r-\varrho)^{m+\alpha}w}{(\alpha+m)!\,\overline{L}^{(m)}[1]}\,dx.$$

Here the operator $\overline{L}_k$ is defined by

$$(7.60)\quad \overline{L}_k[w]=-\frac{m!}{(m-k)!}\overline{L}^{(m-k)}\left[\frac{w}{\overline{L}^{(m)}[1]}\right]\quad \text{for } k=1,\ldots,m.$$

$\overline{L}_k$ is an operator of order k. The coefficients of $\overline{L}_k$ are of class C_{s+m-k} for $x\neq z$ and are formed rationally from the $A_{i_1\ldots i_\beta}(x)$ with $\beta\geqq m-k$ and their derivatives of order $\leqq\beta-m+k$.

Formula (7.59) expresses a spherical integral of u with a kernel that vanishes of order α on the boundary of the sphere in terms of spherical integrals of u with kernels that vanish of higher order on the boundary. We can iterate that formula. For that purpose we introduce operators $\overline{L}_k^j$ for $j=1,2,\ldots,s+1$ and $k=j,j+1,\ldots,m+j-1$ by the formulae [65]

[65] Notice that the summation convention is *not used* here.

$$(7.61a)\qquad \bar{L}_k^1 = \bar{L}_k \qquad \text{for } k = 1, \ldots, m$$

$$(7.61b)\qquad \bar{L}_k^{j+1} = \bar{L}_k^j + \bar{L}_{k-j}\bar{L}_j^j \text{ for } k = j+1, \ldots, m+j-1$$

$$(7.61c)\qquad \bar{L}_{m+j}^{j+1} = \bar{L}_m \bar{L}_j^j.$$

It follows by induction over j that $\bar{L}_k^j$ is an operator of order k with coefficients of class C_{s+m-k} for $x \neq z$, which are formed from the $A_{i_1 \ldots i_\beta}$ with $\beta \geqq m-k$ and their derivatives of order $\leqq \beta - m + k$.

For simplicity we also introduce the abbreviation

$$(7.61d)\qquad \bar{L}_0^0 = 1.$$

Let j be one of the numbers $j = 0, \ldots, s$. Let $w(x)$ be a function of class C_{m+j}, which vanishes in a neighborhood of the point $x = z$. Then the identity

$$(7.62)\qquad \int\limits_{\varrho \leqq r} uw\,dx = \sum_{k=j+1}^{m+j} \int\limits_{\varrho \leqq r} u \frac{(r-\varrho)^k}{k!} \bar{L}_k^{j+1}[w]\,dx$$

$$+\, m! \int\limits_{\varrho < r} \frac{f}{\bar{L}^{(m)}[1]} \sum_{k=0}^{j} \frac{(r-\varrho)^{m+k}}{(m+k)!} \bar{L}_k^k[w]\,dx$$

expresses a weighted integral of the weak solution u in terms of weighted integrals, which vanish of order $j+1$ on the boundary $\varrho = r$. This identity is proved by induction over j. For $j = 0$ we just have formula (7.59) with $\alpha = 0$. Moreover (7.62) can be reduced to the corresponding formula with j replaced by $j-1$ by using the recursion formulae (7.61b, c) and also using the formula (7.59) with α and w replaced respectively by j and by $\bar{L}_j^j[w]$.

We take for $w(x)$ the function defined by

$$(7.62a)\qquad w(x) = 1 - \theta\left(\frac{1}{\varepsilon}\varrho - 1\right) = 1 - \theta\left(\frac{1}{\varepsilon}\,|\,x-z\,| - 1\right),$$

where $\theta(s)$ is again a function of class C_∞ with the properties (7.52a). Then

$$w(x) = 0 \text{ for } |\,x-z\,| < \varepsilon;\ w(z) = 1 \text{ for } |\,x-z\,| > 2\varepsilon.$$

Differentiation of (7.62) with respect to r yields for $r > 2\varepsilon$

$$\omega_n I(z, r) r^{n-1} = \sum_{k=j+1}^{m+j} \int_{\varrho<r} u(x) \frac{(r-\varrho)^{k-1}}{(k-1)!} \overline{L}_k^{j+1}[w]\, dx \tag{7.63}$$

$$+ m! \int_{\varrho<r} \frac{f(x)}{\overline{L}^{(m)}[1]} \sum_{k=0}^{j} \frac{(r-\varrho)^{m+k-1}}{(m+k-1)!} \overline{L}_k^k[w]\, dx,$$

where $I(z, r)$ is the spherical average of $u(x)$ on the sphere of radius r and center z. The right hand side of (7.63) depends on z and r through the domain of integration $|x - z| < r$ and through the integrand. The integrand contains r and z in the powers of $r - \varrho = r - |x - z|$ that occur, and in w, which depends on $\varrho = |x - z|$. Moreover z enters the expression $\overline{L}^{(k)}$ by (7.55a) and thus also enters the coefficients of all $\overline{L}_k^j$ by (7.60), (7.61a, b, c). The integrand is in fact a function of z and r of class C_∞. Looking at one of the terms

$$\int_{\varrho<r} u(x) \frac{(r-\varrho)^{k-1}}{(k-1)!} \overline{L}_k^{j+1}[w]\, dx \tag{7.64}$$

we see that it can be differentiated $k - 1$ times with respect to r or z by differentiating under the integral sign, without any contribution from the boundary. For continuous $u(x)$ one additional differentiation is permitted, which will produce a surface integral as contribution of the variable boundary. The substitution $x = z + r\xi$ will reduce the surface integral to one over the fixed sphere Ω_ξ. The surface integral can then be differentiated as many times with respect to r and z as the integrand has x-derivatives. Since $\overline{L}_k^{j+1}[w]$ is of class C_{s+m-k} in x, z, r, we find that the integral (7.64) is of class C_k for continuous u and of class C_t with $t > k$ for u of class C_{t-k} and $s + m \geqq t$. It follows that the first sum of the right hand side of formula (7.63) is of class C_t for u of class C_{t-j-1} and $s + m \geqq t$, $0 \leqq j \leqq s$. Taking $j = s$ we find that the first sum is of class C_t for u of class C_{t-s-1}. Similarly the second sum on the right hand side of (7.63) is of class C_t for f in C_{t-m} and $t \leqq 2m + s$.

It follows that the spherical means $I(z, r)$ of a continuous weak solution u of (7.50) are of class C_t in z and r (for $r > 2\varepsilon$), if $u(x)$ is of class C_{t-s-1}, f of class C_{t-m} and $s + m \geqq t$.[66]

We now make use of the theorem that a continuous function $u(x) = u(x_1, \ldots, x_n)$ is of class $C_{t-2[n/2]}$, if its spherical means $I(z, r)$ are of class C_t in z and r for $r > 0$. Replacing t by $t + 2[n/2]$, it follows that u is of class C_t provided u is of class $C_{t+2[n/2]-s-1}$, f of class $C_{t+2[n/2]-m}$ and provided that $t \leqq s + m - 2[n/2]$.

For $s = 2[n/2]$, f of class $C_{2[n/2]}$, it follows that u in C_{t-1} has as a consequence u in C_t for $t = 1, \ldots, m$. In this case then continuity of u implies that u is in C_m. We thus have the following theorem:

Any continuous weak solution $u(x)$ of the linear m-th order elliptic equation

$$L[u(x)] = f(x) \tag{7.65}$$

in the domain D is a strict solution of class C_m, provided f is of class $C_{2[n/2]}$ and the coefficients of L multiplying the derivatives of order β are of class $C_{\beta+2[n/2]}$ for $\beta = 0, \ldots, m$.

More generally we have for $s \geqq 2[n/2]$:

A continuous weak solution of the linear m-th order elliptic equation (7.65) is a strict solution of class $C_{m+s-2[n/2]}$ provided f is of class C_s and the coefficients of the derivatives of order β in L are of class $C_{\beta+s}$.

For $s = m - 1$, $t = m - 2[n/2]$, we find:

A continuous weak solution of the linear elliptic equation (7.65) is of class $C_{m-2[n/2]}$ provided $f(x)$ is continuous and the coefficients of L are in C_∞.

As an illustration of the last theorem we have that a continuous function $u(x)$ is of class $C_{2k-2[n/2]}$, if the iterated Laplace operator Δ^k can be applied to u in the weak sense and yields a continuous function (i.e. if u is a weak solution of an equation of the form $\Delta^k u = f(x)$ with a continuous f).

[66] We use the convention that the class C_i for any $i \leqq 0$ consists of all continuous functions.

Explicit representations and estimates for the derivatives of a solution of a linear elliptic equation

We consider the linear elliptic differential equation

$$L[u] = 0 \tag{7.66}$$

of order m. The coefficients $A_{i_1 \ldots i_\beta}(x)$ of the derivatives of order β shall be of class $C_{s+\beta}$ in the domain D, where s is some integer. Let $u(x)$ be a (strict) solution of (7.66) of class C_m in D. We have the identity (7.63) for the spherical mean of $u(x)$, which here becomes for $0 \leqq j \leqq s$

$$I(z, r) = \frac{1}{\omega_n} r^{1-n} \sum_{k=j+1}^{m+j} \int_{\varrho<r} u(x) \frac{(r-\varrho)^{k-1}}{(k-1)!} \overline{L}_k^{j+1}[w]\, dx. \tag{7.67}$$

This identity holds, if an r-neighborhood of z lies in D. The function $w(x)$ is defined by (7.62a) with some $\varepsilon < r/2$. The expression $\overline{L}_k^{j+1}[w]$ is formed rationally from the $A_{i_1 \ldots i_\beta}(x)$ with $\beta \geqq m - k$ and their derivatives of order $\leqq \beta - m + k$, the $x_i - z_i$, $|x - z|$, and the function w and its derivatives. From (7.67) we obtain by differentiation expressions for any derivatives of $I(z, r)$ with respect to z or r of order $\leqq j + 1$. Those expressions will consist of weighted integrals of u over the surface and also over the interior of the sphere $|x - z| = r$. The weights depend on the same $A_{i_1 \ldots i_\beta}(x)$ and their same x-derivatives as those occurring in the expression for I itself.

According to formulae (4.24), (4.26) $u(x)$ can be expressed in terms of the iterated spherical mean $M(x, \lambda, \mu)$ with $\mu > 0$ and of its derivatives of order $\leqq 2[n/2]$.[67] Any derivative of $u(x)$ of some order t, can then be expressed in terms of the derivatives of M of order $\leqq t + 2[n/2]$. We can substitute for the derivatives

[67] For the present purposes it would be sufficient to consider just the case of odd n, where u can be expressed by (4.24) as a linear combination of derivatives of M of orders $\leqq n - 1$. The results on the derivatives of u for odd n that will be obtained can be generalized subsequently for even n by Hadamard's principle of descent: A solution $u(x_1, \ldots, x_n)$ of (7.66) for even n is also a solution of the elliptic equation $L[u] + \partial^m u/\partial x_{n+1}^m = 0$, if the characteristic form of L is positive definite. The latter equation has an odd number of independent variables.

of $M(x, \lambda, \mu)$ their expressions as spherical means of derivatives of $I(z, r)$, and for those in turn their expressions as weighted integrals of u. We thus obtain for $t \leqq j + 1 - 2[n/2]$ an expression for the partial derivatives of u of order t at any point of D in terms of repeated weighted spherical integrals of u in a neighborhood of the point. The weight functions occurring in these integrals contain the $A_{i_1 \ldots i_\beta}(x)$ and their derivatives of orders $\leqq \beta + j$. For $j = t - 1 + 2[n/2]$ we obtain then a *universal* expression for any t-th derivative of u in terms of integrals of u, provided u satisfies some linear elliptic differential equation. The integrands in this expression depend explicitly and rationally on the coefficients $A_{i_1 \ldots i_\beta}(x)$ of L and on their derivatives of orders $\leqq \beta + t - 1 + 2[n/2]$. (The denominators in this rational function are powers of the characteristic form $Q(x, \xi)$ of L.) In addition the expression contains the parameters ε and λ or β (through formulae (4.24), (4.26)), which have to be chosen in accordance with the distance of x from the boundary of D.

From the existence of an expression of this type follow a priori estimates for the derivatives of u at a point $x = z$ in terms of the maximum of the absolute value of u in a neighborhood of z. We first assume that for the points z in question there exists a unit sphere with center z, which is completely contained in D. We then have the following result:

If $u(x)$ is a solution of a linear elliptic differential equation $L[u] = 0$ in a unit sphere about the point z, then any derivative of u of order t has at the point z at most the value

$$M(n, m, t, N) \left(\underset{|x-z| \leqq 1}{\text{Maximum}} \, |u(x)|\right). \tag{7.68}$$

Here $M(n, m, t, N)$ is a universal function of its arguments. The order of the equation is m, the number of independent variables is n. The letter N stands for a common upper bound of

a) $\underset{\substack{|x-z| \leqq 1 \\ |\xi| = 1}}{\text{Maximum}} \, |1/Q(x, \xi)|$, where $Q(x, \xi)$ is the characteristic form of L, and

b) the maximum in $|x - z| \leqq 1$ of the absolute values of the derivatives of orders $\leqq \beta + t - 1 + 2[n/2]$ of the coefficients $A_{i_1 \ldots i_\beta}(x)$ of the terms of order β occurring in the differential operator L.

A similar result can be obtained in points that have a smaller distance from the boundary Let $u(x)$ be a solution of a linear elliptic equation $L[u] = 0$ in a sphere of radius d about a point z. Let again m, n and $A_{i_1 \ldots i_\beta}(x)$ be defined as before. Apply the substitution $x = z + d(x' - z)$. Then $u'(x') = u(z + d(x' - z))$ is a solution of a linear elliptic equation $L'[u'(x')] = d^m L[u] = 0$ in a unit sphere about the point z of x'-space. The coefficients of terms of order β in the operator L' are given by

$$A'_{i_1 \ldots i_\beta}(x') = d^{m-\beta} A_{i_1 \ldots i_\beta}(z + d(x' - z)); \quad (\beta = 0, 1, \ldots, m).$$

For $d \leqq 1$ the derivatives of $A'_{i_1 \ldots i_\beta}(x')$ in the sphere $|x' - z| \leqq 1$ do not exceed the values of the corresponding derivatives of the $A_{i_1 \ldots i_\beta}(x)$ in the sphere $|x - z| \leqq d$. The characteristic forms of L and L' are the same in corresponding points. A derivative of $u'(x')$ of order t differs by the factor d^t from the derivative of $u(x)$ at the corresponding point. Hence:

If $u(x)$ is a solution of the linear elliptic equation $L[u] = 0$ in a sphere of radius $d \leqq 1$ about the point z, then any derivative of u of order t at the point z has at most the value

$$d^{-t} M(n, m, t, N) \underset{|x-z| \leqq d}{\text{Maximum}} |u(x)|. \tag{7.69}$$

Here $M(n, m, t, N)$ is the same universal function as before. For N we take a common upper bound of

a) $$\underset{\substack{|x-z| \leqq d \\ |\xi|=1}}{\text{Maximum}} |1/Q(x, \xi)|$$

and of

b) the maximum in $|x - z| \leqq d$ of the absolute values of the derivatives of order $\leqq \beta + t - 1 + 2[n/2]$ of the $A_{i_1 \ldots i_\beta}(x)$.

Let L be a linear elliptic operator with coefficients $A_{i_1 \ldots i_\beta}(x)$ of class $C_{s+\beta}$ in a domain D. Then for any family of solutions u of $L[u] = 0$ in D, for which the u have the same common bound in D, *there exists in every compact subset of D a common bound for all derivatives of order t of those u, provided* $t \leqq s + 1 - 2[n/2]$.

Let L be a linear elliptic operator with analytic coefficients in a domain D. Let $u(x)$ be a solution of $L[u] = 0$ with an isolated singularity at a point y of D. We saw that then $u(x)$ is analytic for $x \neq y$. We also proved in Chapter III, p. 58, that $u(x)$ is of the form $u(x) = cK(x, y) + w(x)$, where c is a constant, $K(x, y)$ a fundamental solution with pole y, and $w(x)$ a regular analytic solution at $x = y$, provided the $(m - 1)$-st derivatives of $u(x)$ multiplied by $|x - y|^n$ tend to 0 for $x \to y$. We are now in a position to weaken the assumptions. It follows indeed from (7.69) that any t-th derivative of u multiplied with $|x - y|^t$ is bounded near $x = y$, if u itself is bounded near $x = y$. For $t = m - 1$ and $m < n + 1$ we then have the result:

If $u(x)$ is a solution of $L[u] = 0$, which has an isolated singularity at $x = y$ and is bounded near $x = y$, then for $m < n + 1$, u is of the form

$$u(x) = cK(x, y) + w(x), \tag{7.70}$$

where c is a constant, K a fundamental solution and w regular at $x = y$.

We saw in Chapter III, p. 61, that $K(x, y)$ is of the form (3.43). For $m = n$ the logarithmic term is certainly present, as was shown on p. 65, and hence $K(x, y)$ is certainly not bounded for x near y. For $m < n$ the fundamental solution is again not bounded for x near y, since otherwise the m-th order derivatives of $K(x, y)$ would be integrable over volumes, contrary to the basic properties of the fundamental solution. Hence we have the more precise information:

For a linear elliptic differential equation with analytic coefficients and an order not exceeding the number of dimensions boundedness of a solution in a neighborhood of an isolated singularity implies regularity of the solution. [67a]

One proves in the same way for linear elliptic differential equations with analytic coefficients and of order $m \geqq n + 1$ that u is of the form (7.70), if $\lim_{x \to y} |x - y|^{n-m+1} u(x) = 0$.

[67a] See Miranda [1], p. 52 for related results with $m = 2$.

CHAPTER VIII

Regularity Properties for Integrals of Solutions over Time-like Lines

The methods used in Chapter VII for establishing differentiability of solutions of elliptic equations can be applied to some extent to non-elliptic equations as well. One obtains then as a result differentiability properties not of the solution itself but of certain of its integrals.[68]

Definition of "time-like"

We consider a linear differential operator L of order m in the $n+1$ independent variables $x_1, \ldots, x_n, t$. Let $Q = Q(x, t, \xi, \tau)$ be the characteristic form of L, where $x = (x_1, \ldots, x_n)$, $\xi = (\xi_1, \ldots, \xi_n)$. A manifold M in $(n+1)$-dimensional xt-space is called *free* at one of its points P, if $Q \neq 0$ for the direction numbers $(\xi_1, \ldots, \xi_n, \tau)$ of every normal of M at P. If M is an n-dimensional hyper surface, freedom of M is equivalent to not being characteristic. At the other extreme we can consider one-dimensional curves M. A free curve will be called *time-like*. We shall assume in this chapter that L is such that all lines parallel to the t-axis are time-like. This amounts to the assumption that

$$(8.1) \qquad Q(x, t, \xi, 0) \neq 0$$

for all $\xi \neq 0$ and all x, t in question. In the terminology of Chapter II condition (8.1) expresses the fact that the normal surface of L is bounded.

The corresponding canonical system

As explained in Chapter VII, p. 130, the equation

$$(8.2) \qquad L[u] = 0$$

[68] See John [10].

can be reduced to a linear canonical system of equations for the dependent variables $u_1(x, t), \ldots, u_N(x, t)$, which stand for $u(x, t)$ and its partial derivatives of order $\leqq m - 1$. These canonical equations express any first derivative of a $u_i(x, t)$ linearly in terms of the u_k and of first derivatives of the u_k in a direction perpendicular to (ξ, τ). The coefficients of the canonical equations are formed rationally from the coefficients of L and the quantities ξ, τ. The only denominator occurring is the characteristic form $Q(x, t, \xi, \tau)$ and $|\xi|^2 + \tau^2$. It follows from (8.1) that for $\tau = 0$, $\xi \neq 0$ the coefficients of the canonical system are regular functions of x, t, ξ with as many continuous derivatives as the coefficients of L have.

For $\tau = 0$ the canonical system contains equations of the form

$$(8.3) \quad u_{k,\alpha}(x, t) = S_{ks\alpha\beta}(x, t, \xi)u_{s,\beta}(x, t) + R_{ks\alpha}(x, t, \xi)u_{s,t}(x, t) + T_{ki\alpha}(x, t, \xi)u_i(x, t)$$
$$(k, s = 1, \ldots, N;\ \alpha, \beta = 1, \ldots, n).$$

Here the summation convention is used. Moreover $u_{s,\beta}(x, t)$ denotes the partial derivative of $u_s(x, t)$ with respect to x_β, and $u_{s,t}(x, t)$ that with respect to t. The $S_{ms\alpha\beta}$, $R_{ms\alpha}$, $T_{mi\alpha}$ are rational in ξ with coefficients that are of class C_s in x and t, if the coefficients of L are of that class. The only real pole occurs for $\xi = 0$. The $S_{ms\alpha\beta}(x, t, \xi)$ satisfy the identity

$$(8.4) \qquad S_{ms\alpha\beta}(x, t, \xi)\xi_\beta = 0.$$

Derivatives of cylindrical integrals of a solution

Let the functions $a_i(x, t, \xi)$ for $i = 1, \ldots, N$ be of class C_1 in their arguments for $\xi \neq 0$. We form the expression

$$(8.5) \qquad J(z, r, T) = \int_{-T}^{+T} \left(\int_{|x-z|=r} a_i(x, t, x - z)u_i(x, t)\, dS_x \right) dt.$$

We can form the derivatives of J with respect to z_α and r exactly like in formulae (7.31a, b) the only difference being that the canonical system (8.3) leads to integrals over the cylinder in xt-

space that will also contain t-derivatives of the u_i. Those can be removed by integration by parts with respect to t. In this manner one obtains the formulae

$$\frac{\partial J(z, r, T)}{\partial z_\alpha} = \int_{-T}^{+T} \left(\int_{|x-z|=r} a_i'(x, t, x-z) u_i(x, t) dS_x \right) dt \tag{8.6a}$$

$$\frac{\partial J(z, r, T)}{\partial r} = \int_{-T}^{+T} \left(\int_{|x-z|=r} a_i''(x, t, x-z) u_i(x, t) dS_x \right) dt \tag{8.6b}$$

provided that

$$a_i(x, t, \xi) = 0 \text{ for } t = \pm T \text{ and all } i, x, \xi. \tag{8.7}$$

The $a_i'(x, t, \xi)$ and $a_i''(x, t, \xi)$ are here given by the formulae [see (7.32a, c)]

$$\begin{aligned} a_i'(x, t, \xi) = {} & a_{i,\alpha} - a_k S_{ki\alpha\beta,\beta} - a_k S_{ki\alpha\beta;\beta} - a_{k,\alpha} S_{ki\alpha\beta} \\ & - a_{k;\alpha} S_{ki\alpha\beta} - a_{k,t} R_{ki\alpha} - a_k R_{ki\alpha,t} + a_k T_{ki\alpha} \end{aligned} \tag{8.8a}$$

$$\begin{aligned} |\xi| a_i''(x, t, \xi) = {} & a_{i,\alpha}\xi_\alpha + a_{i;\alpha}\xi_\alpha + (n-1)a_i - a_{k,\beta} S_{ki\alpha\beta}\xi_\alpha \\ & - a_{k;\beta} S_{ki\alpha\beta}\xi_\alpha - a_k S_{ki\alpha\beta,\beta}\xi_\alpha - a_k S_{ki\alpha\beta;\beta}\xi_\alpha \\ & - a_k S_{ki\alpha\alpha} + a_k T_{ki\alpha}\xi_\alpha - a_{k,t} R_{ki\alpha}\xi_\alpha - a_k R_{ki\alpha,t}\xi_\alpha . \end{aligned} \tag{8.8b}$$

(The notation $a_{i;\alpha}$ is used again for the partial derivative of $a_i(x, t, \xi)$ with respect to ξ_α).

If not only the a_i but also the $a_{i,t}$ vanish for $t = \pm T$ we can iterate the process and express the second derivatives of J in terms of integrals of the u_i over the cylinder $|x - z| = r$, $|t| < T$ in xt-space.

Differentiability of integrals of solutions over time-like curves

Let the $a_i(x, t, \xi)$ vanish of order k for $t = \pm T$. If the coefficients of L are all of class C_k we can iterate formulae (8.6a, b) k times. We find that for a solution $u(x, t)$ of (8.2) of class C_m

the cylindrical integrals $J(z, r, T)$ are of class C_k in z and r for $r > 0$. Take in particular for the a_i the expressions

$$a_i(x, t, \xi) = \frac{1}{\omega_n} | \xi |^{1-n} (T^2 - t^2)^k \delta_i^j w(x, t), \tag{8.9}$$

where w is some function of class C_∞. It follows that the expression

$$J(z, r, T) = \frac{r^{1-n}}{\omega_n} \int\limits_{|x-z|=r} \left(\int\limits_{-T}^{+T} w(x, t)(T^2-t^2)^k u_j(x, t)dt \right) dS_x \tag{8.10}$$

is of class C_k in z and r for $r > 0$. The expression $J(z, r, T)$ is the average on a sphere of radius r and center z for the function

$$v_j(x, T) = \int\limits_{-T}^{+T} w(x, t)(T^2 - t^2)^k u_j(x, t)dt \tag{8.11}$$

considered as a function of x. It follows from the theorems of Chapter IV, that $v_j(x, T)$ itself is a function of x of class $C_{k-2[n/2]}$.

The $u_j(x, t)$ include all derivatives of order $\leqq m - 1$ of the solution $u(x, t)$ of (8.2). Let

$$v(x, T) = \int\limits_{-T}^{+T} (T^2 - t^2)^k w(x, t) u(x, t)\, dt. \tag{8.12}$$

The x-derivatives of $v(x, T)$ of orders $\leqq m - 1$ are then linear combinations of functions $v_j(x, T)$ (formed with different weight functions w). Thus $v(x, T)$ is of class $C_{k+m-1-2[n/2]}$ in x. Since differentiation of $v(x, T)$ with respect to T leads to an expression of the same type, only with k replaced by $k - 1$ and since u is of class C_m, we find that $v(x, T)$ is of class $C_{k+m-1-2[n/2]}$ in x *and* T. Consequently we have the theorem:

Let $u(x, t)$ be a solution of class C_m of the m-th order linear differential equation $L[u] = 0$. Let the coefficients of L be of class C_k and let the parallels to the t-axis be time-like with respect to L.

Then for any positive integer k and any $w(x, t)$ of class C_∞ the expression (8.12) *is of class $C_{k+m-1-2[n/2]}$ in x and T.*

The theorem can be brought into a form, where it states that a certain succession of differential and integration processes can be applied indefinitely to $u(x, t)$, provided the coefficients of L are sufficiently often differentiable. We introduce the integral operator σ, which is to transform functions $\Phi(t)$ into functions $\sigma[\Phi(t)]$, and is defined by the formula

$$(8.13) \qquad 2t\int_0^t \Phi(\tau)\,d\tau = \sigma[\Phi].$$

Symbolically we can write

$$(8.14) \qquad \sigma = 2t(d/dt)^{-1}.$$

Then

$$(8.15) \qquad \sigma^k[u(x, t) + u(x, -t)] = \frac{2t}{(k-1)!}\int_{-t}^{+t} (t^2-\tau^2)^{k-1}u(x, \tau)\,d\tau.$$

Our theorem states that this expression is a function of class $C_{k+m-2-2[n/2]}$. If $m-2-2[n/2]$ is non-negative we have the simple result that $\sigma^k[u(x, t) + u(x, -t)]$ is of class C_k. If $m-2-2[n/2]$ is negative, we observe that u is also a solution of the equation

$$L'[u] = (\Delta_x)^{1+[(n-m)/2]}L[u] = 0,$$

which is of order

$$m' = 2 + 2[n/2]$$

(m is necessarily even). The parallels to the t-axis are time-like also for the operator L'. Since $m'-2-2[n/2] = 0$ the expression $\sigma^k[u(x, t) + u(x, -t)]$ is of class C_k, provided the coefficients of L are sufficiently often differentiable, and provided u is of class $C_{m'}$. For a function of class C_1 the operations σ and $\partial/\partial x_\alpha$ are commutative. Combining these remarks we have the following theorem:

Let $u(x, t)$ be a solution of the m-th order equation $L[u] = 0$ of class C_ν, where

$$\nu = \text{maximum } (m,\ 2 + 2[n/2]). \tag{8.15a}$$

Let L have coefficients of class C_∞ and let the t-lines be time-like with respect to L. Then the operators

$$\frac{\partial}{\partial x_\alpha}\sigma = 2t\frac{\partial}{\partial x_\alpha}\left(\frac{\partial}{\partial t}\right)^{-1}$$

($\alpha = 1, \ldots, n$) can be applied arbitrarily often to the function $u(x, t) + u(x, -t)$.

One can easily verify this theorem directly for the solutions of the wave equation

$$L[u] = \frac{\partial^2}{\partial t^2}u - \Delta_x u = 0. \tag{8.16}$$

If $u(x, t)$ is a solution of (8.16) with initial data

$$u = F(x),\ u_t = G(x) \text{ for } t = 0,$$

we have in

$$\begin{aligned} v &= \left(2x_\alpha + \frac{\partial}{\partial x_\alpha}\sigma\right)[u(x, t) + u(x, -t)] \\ &= 2x_\alpha(u(x, t) + u(x, -t)) + 2t\frac{\partial}{\partial x_\alpha}\int_{-t}^{+t} u(x, \tau)\,d\tau \end{aligned} \tag{8.17}$$

the solution of the wave equation with initial data

$$v = 4x_\alpha F(x),\ v_t = 0 \text{ for } t = 0.$$

Since a solution of class C_2 of the wave equation exists for initial data of class $C_{[(n+4)/2]}$ it is clear that the operators $(2x_\alpha + (\partial/\partial x_\alpha)\,\sigma)$ can be applied arbitrarily often to a solution of the wave equation with initial data $u = F(x)$, $u_t = 0$, where F is of class $C_{[(n+4)/2]}$. The same holds then for the operators $(\partial/\partial x_\alpha)\sigma$ alone.

Integrals of solutions over time-like curves with common endpoints

Instead of considering weighted integrals of a solution u over *parallel* line segments and with weights vanishing of a certain order in the endpoints, we can also take integrals over a family of time-like lines *with common endpoints*.

Let $\Phi(t)$ be a function of class C_∞ with the property that

$$\Phi(t) = 0 \text{ for } t = \pm T. \tag{8.18}$$

We consider the expression

$$w(x, \xi) = \int_{-T}^{+T} u(x + \xi\Phi(t), t)\, dt. \tag{8.19}$$

For any x, ξ this is an integral of the function u over a curve $C_{x,\xi}$ with endpoints $(x, \pm T)$, which are independent of ξ. For $\xi = 0$ these curves reduce to parallels to the t-axis, and hence are time-like. It is evident then that the $C_{x,\xi}$ will also be time-like for all sufficiently small ξ.

We have

$$\frac{\partial w}{\partial \xi_\alpha} = \frac{\partial}{\partial x_\alpha} \int_{-T}^{+T} u(x + \xi\Phi(t), t)\, \Phi(t)\, dt = \frac{\partial}{\partial x_\alpha} \int_{-T}^{+T} u'(x, t)\, \Phi(t)\, dt. \tag{8.20}$$

Here $u'(x, t)$ for fixed ξ is defined by $u(x + \xi\Phi(t), t)$. Similarly any ξ-derivative of w of order k can be represented as an x-derivative of order k of the expression.

$$\int_{-T}^{+T} u'(x, t)\, \Phi^k(t) dt. \tag{8.21}$$

Now $u'(x, t)$ is a solution of a linear equation $L'[u'] = 0$, where L' depends also on the parameter ξ and is close to L for small ξ. Moreover the weight $\Phi^k(t)$ in the integral (8.21) vanishes of order k for $t = \pm T$. It follows then that for L with coefficients in C_∞ and for u of class C_ν (where ν is defined by (8.15a)) the expression (8.21) is of class C_k in x. This suggests that all ξ-derivatives of $w(x, \xi)$ of order k exists, since they can be expressed formally in terms of expressions which are known to exist. This argument for the existence of all ξ-derivatives of the expression (8.19) can easily be converted into a rigorous proof, using induction over the number of derivatives involved. One only has to recall that the x-derivatives of order k of the expression (8.21) can actually be expressed explicitly as weighted integrals of u and its derivatives of order $\leqq \nu - 1$. One further ξ-differentiation can then always be carried out under the integral sign, if u is of class C_ν. Hence, putting $x = 0$:

Given an operator L with coefficients of class C_∞ and a family of curves of class C_∞ and parameters ξ of the form

$$x = \xi \Phi(t) \quad \text{for} \quad |t| \leqq T \tag{8.22}$$

with common endpoints for $t = \pm T$, and a solution $u(x, t)$ of $L[u] = 0$ of class C_ν, the integral of u with respect to t over those curves is a function of ξ of class C_∞.

In case $\Phi(t)$ and the coefficients of L are analytic functions, one would expect the integrals of u over the curves (8.22) to be analytic functions of ξ. It appears plausible that a proof [69] could be given by suitably estimating the ξ-derivatives of $w(x, \xi)$, whose existence has been established, similar to the analyticity proof for solutions of linear analytic elliptic equations on pp. 142 et seq.

[69] A quite different proof of this fact, which uses solutions of special Cauchy problems for the adjoint equation is given in John [4], p. 248.

BIBLIOGRAPHY

ASGEIRSSON, L.

[1] Über eine Mittelwertseigenschaft von Lösungen homogener linearer partieller Differentialgleichungen. 2. Ordnung mit konstanten Koeffizienten. Math. Ann. 113 (1937), pp. 321—346.

BERS, L.

[1] Local behavior of solutions of general linear elliptic equations. Communications on Pure and Applied Mathematics 8 (1955), no. 4.

BROWDER, FELIX E.

[1] Assumption of boundary values and the Green's function in the Dirichlet problem for the general linear elliptic equation. Proc. Nat. Acad. Sci. U.S.A. 39 (1953), pp. 179—184.

[2] The Dirichlet and vibration problems for linear elliptic differential equations of arbitrary order. Proc. Nat. Acad. Sci. U.S.A. 38 (1952), pp. 741—747.

[3] The Dirichlet problem for linear elliptic equations of arbitrary even order with variable coefficients. Proc. Nat. Acad. Sci. U.S.A. 38 (1952), pp. 230—235.

[4] Strongly elliptic systems of differential equations. Contributions to the theory of partial differential equations. Ann. Math. Studies 33 (1954), pp. 15—51.

BRUSOTTI, L.

[1] Sopra alcune questioni di geometria suggerite dalla teoria delle equazioni a derivate parziali totalmente iperboliche. Acad. Roy. Belgique, Bull. Cl. Sci. (5) 39 (1953), pp. 381—404.

BUREAU, FLORENT

[1] Quelques questions de Géométrie suggérées par la théorie des équations aus dérivées partielles totalement hyperboliques. Colloque de Géométrie Algébrique, Liège, 1949.

[2] Essai sur l'intégration des équations linéaires aux dérivées partielles. Mém. Acad. roy. Belgique, Cl. Sci., 2e sér. v. XV, 1936.

[3] Sur l'intégration des équations linéaires aux dérivées partielles. Acad. Roy. Belgique, Bull. Cl. Sci. (5) 22 (1936), pp. 156—174.

[4] Les solutions élementaires des équations linéaires aux dérivées partielles. Bruxelles (Marcel Hayez), 1936.

[5] Le problème de Cauchy pour une équation linéaire aux dérivées partielles, totalement hyperbolique, d'ordre et à quatre variables indépendantes. Acad. Roy. Belgique, Bull. Cl. Sci. (5) 33 (1947), pp. 379—402.

[6] Sur la solution élémentaire d'une équation linéaire aux dérivées partielles d'ordres quatre et à trois variables indépendantes. Acad. Roy. Belgique, Bull. Cl. Sci. (5) 33 (1947), pp. 473—484.

[7] La solution élémentaire d'une équation linéaire aux dérivées partielles, décomposable et totalement hyperboliqeu, d'ordre quatre et à quatre variables indépendantes. Acad. Roy. Belgique, Bull. Cl. Sci. (5) 34 (1938), pp. 566—592.
[8] Intégrales de Fourier et problème de Cauchy. Anali di Matematica pura ed applicata, ser. 4, 32 (1951), pp. 205—233.
[9] Divergent integrals and partial differential equations. U. of Chicago, 1954 (work sponsored by the office of Ordnance Research under contract DA 11-022-ORD-1318).

CAUCHY, A.
[1] Mémoire sur l'intégration des équations linéaires aux differentielles partielles et à coefficients constants. Oeuvres Completes, 2e série, v. I, pp. 275—357.

COOPER, J. L. B.
[1] The application of multiple Fourier transforms to the solution of partial differential equations. Quart. J. Math., Oxford, Ser. (2) 1 (1950), pp. 122—135.

COURANT, R., and HILBERT, D.
[1] Methoden der mathematischen Physik. Springer, Berlin, 1937.
[2] Methods of Mathematical Physics, v. II Interscience, New York-London, in preparation.

COURANT, R., and LAX, A.
[1] Remarks on Cauchy's problem for hyperbolic partial differential equations with constant coefficients in several independent variables. Communications on Pure and Applied Mathematics 8 (1955), no. 4.

DELSARTE, J.
[1] Les fonctions "moyenne-periodiques." J. Math. Pures Appl., sér. 14, 9 (1935), pp. 409—453.

DIAZ, J. B., and WEINBERGER, H. F.
[1] A solution of the singular initial value problem for the Euler-Poisson-Darboux equation. Proc. Am. Math. Soc. 4 (1953), pp. 703—715.

EHRENPREIS, LEON
[1] Solution of some problems of division. Amer. J. Math. 76 (1954), pp. 883—903.

FRIEDRICHS, K. O.
[1] On the differentiability of the solutions of linear elliptic differential equations. Communications on Pure and Applied Mathematics 6 (1953), pp. 299—326.
[2] On differential operators in Hilbert spaces. Amer. J. Math. 61 (1939), pp. 523—544.

FREDHOLM, I.
[1] Sur l'intégrale fondamentale d'une équation differentielle elliptique à coefficients constants. Rend. Circ. Mat. Palermo 25 (1908), pp. 346—351.

GÅRDING, L.
[1] Linear hyperbolic partial differential equations with constant coefficients. Acta Math. 85 (1950), pp. 1—62.

[2] On a lemma by H. Weyl. Kungl. Fysiografiska Sällakapets i Lund Förhandlingar 20 (1950), no. 23, pp. 1—4.

HADAMARD, J.
[1] Lectures on Cauchy's problem in linear partial differential equations. Reprinted by Dover, New York, 1952.

HERGLOTZ, G.
[1] Über die Integration linearer partieller Differentialgleichungen mit konstanten Koeffizienten. Ber. Math. Phys. Kl. Sächs. Akad. Wiss., Leipzig. Part I (Anwendung Abelscher Integrale), 78 (1926), pp. 41—74; Part II (Anwendung Fourierscher Integrale), 78 (1926), pp. 287—318; Part III (Anwendung), 80 (1928), pp. 69—114.
[2] Über die Integration linearer partieller Differentialgleichungen mit konstanten Koeffizienten. Abh. Math. Sem. Hamburg (1928), pp. 189—197.
[3] Notes of Lectures on "Mechanik der Kontinua," Göttingen, 1931.

HOPF, E.
[1] Über den funktionalen, insbesondere den analytischen Charakter der Lösungen elliptischer Differentialgleichungen zweiter Ordnung. Math. Zeit. 34 (1931), pp. 191—233.

HOWARD, AUGHTUM S.
[1] Linear second order partial differential equations with constant coefficients. Dissertation, Graduate School, University of Kentucky, 1942.

JOHN, FRITZ
[1] Bestimmung einer Funktion aus ihren Integralen über gewisse Mannigfaltigkeiten, Math. Ann. 109 (1934), pp. 488—520.
[2] Abhängigkeiten zwischen den Flächenintegralen einer stetigen Funktion. Math. Ann. 111 (1935), pp. 541—559.
[3] Linear partial differential equations with analytic coefficients. Proc. Nat. Acad. Sci. U.S.A. 29 (1943), pp. 98—104.
[4] On linear partial differential equations with analytic coefficients. (Unique continuation of data). Communications on Pure and Applied Mathematics 2 (1949), pp. 209—253.
[5] The fundamental solution of linear elliptic differential equations with analytic coefficients. Communications on Pure and Applied Mathematics 3 (1950), pp. 273—304.
[6] The ultrahyperbolic differential equation with 4 independent variables. Duke Math. J. 4 (1938), pp. 300—322.
[7] General properties of solutions of linear elliptic partial differential equations, Proceedings of the Symposium on Spectral Theory and Differential Problems, Stillwater, Oklahoma (1951), pp. 113—175.
[8] Special topics in Partial Differential Equations, Lectures given in the spring of 1952 at New York University. Notes by M. Jordan, Institute for Mathematics and Mechanics.
[9] Derivatives of continuous weak solutions of linear elliptic equations. Communications on Pure and Applied Mathematics 6 (1953), pp. 327—335.

[10] On behavior of solutions of partial differential equations. University of Maryland, Institute for Fluid Dynamics, Lecture Series No. 25 (1953).

KELLOG, O. D.
[1] Foundations of potential theory, Springer, Berlin, 1929.

LAX, P. D.
[1] On Cauchy's problem for hyperbolic equations and the differentiability of solutions of elliptic equations. Communications on Pure and Applied Mathematics 8 (1955), no. 4.

LERAY, JEAN
[1] Notes of lectures on "Symbolic calculus with several variables, projections and boundary value problems for differential equations." Princeton University, 1951.
[2] On linear hyperbolic equations with variable coefficients in a vector space. Ann. Math. Studies 33 (1954), pp. 201—210.

LEVI, E. E.
[1] Sulle equazioni lineari totalmente ellittiche alle derivate parziali. Rend. del Circ. Mat. Palermo 24 (1907), pp. 275—317.

LOPATINSKII, YA. B.
[1] A fundamental system of solutions of a system of linear differential equations of elliptic type. C. R. (Doklady) Acad. Sci. URSS (N.S.) 71 (1950), pp. 433—436.
[2] The normal fundamental solutions of a system of linear differential equations of elliptic type. C. R. (Doklady) Acad. Sci. URSS (N.S.) 78 (1951), pp. 865—867.

MADER, PHILOMENA
[1] Über die Darstellung von Punktfunktionen im n-dimensionalen euklidischen Raum durch Ebenenintegrale. Math. Zeit. 26 (1927), pp. 646—652.

MAGNUS, W., and OBERHETTINGER, F.
[1] Formulas and theorems for the special functions of mathematical physics. Chelsea Publ. Co., 1949.

MALGRANGE, B.
[1] Equations aux dérivées partielles à coefficients constants. 1. Solutions élémentaire. C. R. Acad. Sci. Paris. 237 (1953), pp. 1620—1622. 2. Equations avec second membre. Ibid. 238 (1954), pp. 196—198.

MIRANDA, C.
[1] Equazioni alle derivate parziale di tipo ellittico. Ergebnisse der Mathematik, Neue Folge, Heft 2 (1955).

MORREY, C. B.
[1] On the solutions of quasilinear elliptic partial differential equations. Trans. Amer. Math. Soc. 43 (1938), pp. 126—166.
[2] Second order elliptic systems of differential equations. Proc. Nat. Acad. Sci. U.S.A. 39 (1953), pp. 201—6.
[3] Second order elliptic system of differential equations. Contributions to the theory of partial differential equations. Ann. Math. Studies. 33 (1954), pp. 101—159.

MYSKIS, A.
[1] The uniqueness of the solution of Cauchy's problem. Uspekhi Matem. Nauk (N.S.) 3, no. 2 (24), (1948), pp. 3—46.
[2] The normal fundamental solutions of a system of linear differential equations of elliptic type. C. R. (Doklady) Acad. Sci. USSR (N.S.) 78 (1951), pp. 865—867.

NIRENBERG, L.
[1] On non-linear elliptic partial differential equations and Hölder continuity. Communications on Pure and Applied Mathematics 6 (1953), pp. 103—155.
[2] On a generalization of quasi-conformal mappings and its application to elliptic partial differential equations. Contributions to the theory of partial differential equations. Ann. Math. Studies. 33 (1954), pp. 95—100.

PETROVSKII, I.
[1] On the diffusion of waves and the lacunas for hyperbolic equations. Rec. Math. 17 (59), no. 3 (1945).
[2] Sur les systèmes d'équations différentielles dont toutes les solutions sont analytiques. C. R. (Doklady) Acad. Sci. URSS 17, no. 7 (1937), pp. 343—6.
[3] Sur l'analyticite des solutions des systemes d'equations differentielles. Matem. Sbornik (Recueil Mathematique) N.S. 5 (47), v. 1 (1939), pp. 3—70.
[4] On some problems in the theory of partial differential equations. Uspehi Matem. Nauk. (N.S.), I, no. 3—4 (13—14) (1946), pp. 44—70. Amer. Math. Soc. Translation No. 12.

RADON, J.
[1] Über die Bestimmung von Funktionen ducrh ihre Integralwerte längs gewisser Mannigfaltigkieten. Ber. Verh. Sächs. Akad. 69 (1917), pp. 262—277.
RIESZ, MARCEL
[1] L'intégrale de Riemann-Liouville et le problèmė de Cauchy. Acta Math. 81 (1951), pp. 1—223.

SCHWARTZ, LAURENT
[1] Théorie générale des fonctions moyenne-périodiques. Ann. of Math. 48 (1947), pp. 857—929.
[2] Théorie des distributions, Hermann, Paris, 1950.
SERRIN, J. B.
[1] A note on the wave equation. Proc. Amer. Math. Soc. 5 (1954), pp. 307—8.
SHAPIRO, Z.
[1] On elliptical systems of partial differential equations C. R. (Doklady) Acad. Sci. URSS (N. S.) 46 (1945).
SOMIGLIANA, C.
[1] Sui sistemi simmetrici di equazioni a derivati parziali. Annali di Matematica pura ed applicata, ser. 2, 22 (1894), pp. 143—156.

STELLMACHER, KARL L.
[1] Ein Beispiel einer Huggensschen Differentialgleichung. Nachr. Akad. Wiss. Göttingen, Math.-Phys. Kl. No. 10 (1953), pp. 133—138.

TEDONE, O.
[1] Sull' integrazionne dell'equazione $\partial^2\Phi/\partial t^2 - \Sigma_1^m \partial^2\Phi/\partial x_i^2 = 0$. Ann. di Mat. (3), 1 (1898), pp. 1—24.
THOMAS, T. Y., and TITT, E. W.
[1] On the elementary solution of the general linear differential equation of the second order with analytic coefficients. J. de Math. 18 (1939), pp. 217—248.

VISIK, M. I.
[1] The method of orthogonal and direct decomposition in the theory of elliptic differential equations. Mat. Sbornik N.S. 25 (67) pp. 189—234 (1949).
VOLTERRA, V.
[1] Leçons sur les équations intégrales, Paris, 1913.

WEINSTEIN, A.
[1] The Cauchy problem for the wave equation and the equation of Euler-Poisson-Darboux. Institute for Fluid Dynamics and Applied Mathematics, Technical Note BN-22 (1954).
WEYL, H.
[1] The method of orthogonal projection in potential theory. Duke Math. J. 7 (1940), pp. 411—444.

INDEX